Networks and
Distributed Computation

MIT Press Series in Computer Systems
Herb Schwetman, editor

Metamodeling: A Study of Approximations in Queuing Models, by Subhash Chandra Agrawal, 1985

Logic Testing and Design for Testability, by Hideo Fujiwara, 1985

Performance and Evaluation of Lisp Systems, by Richard P. Gabriel, 1985

The LOCUS Distributed System Architecture, edited by Gerald Popek and Bruce J. Walker, 1985

Analysis of Polling Systems, by Hideaki Takagi, 1986

Performance Analysis of Multiple Access Protocols, by Shuji Tasaka, 1986

Performance Models of Multiprocessor Systems, by M. Ajmone Marsan, G. Balbo, and G. Conte, 1986

Microprogrammable Parallel Computer: MUNAP and Its Applications, by Takanobu Baba, 1987

Simulating Computer Systems: Techniques and Tools, by M.H. MacDougall, 1987

A Commonsense Approach to the Theory of Error Correcting Codes, by Benjamin Arazi, 1987

Networks and Distributed Computation: Concepts, Tools, and Algorithms, by Michel Raynal, 1988

Networks and Distributed Computation

Concepts, tools and algorithms

Michel Raynal

Translated by
Meg Sanders

The MIT Press
Cambridge, Massachusetts

First MIT Press edition, 1988

English translation © 1987 North Oxford Academic
Publishers Ltd.

English edition first published in 1987 by North
Oxford Academic Publishers Ltd., a subsidiary of
Kogan Page Ltd., 120 Pentonville Road,
London N1 9JN

Original French language edition, *Systèmes répartis et
réseaux*, © 1987 Éditions Eyrolles.

Printed in Great Britain.

Library of Congress Cataloging-in-Publication Data

Raynal, M.
  Networks and distributed computation.
  (MIT Press series in computer systems)
  Translation of: Systèmes répartis et réseaux.
  Bibliography: p.
  Includes index.
  1. Electronic data processing—Distributed
processing. 2. Computer networks. I. Title.
II. Series.
QA76.9.D5R39   1988   004'.36   87-22842
ISBN 0-262-18130-4

6491

Contents

Series Foreword

This series is devoted to all aspects of computer systems. This means that subjects ranging from circuit components and microprocessors to architecture to supercomputers and systems programming will be appropriate. Analysis of systems will be important as well. System theories are developing, theories that permit deeper understandings of complex interrelationships and their effects on performance, reliability, and usefulness.

We expect to offer books that not only develop new material but also describe projects and systems. In addition to understanding concepts, we need to benefit from the decision making that goes into actual development projects; selection from various alternatives can be crucial to success. We are soliciting contributions in which several aspects of systems are classified and compared. A better understanding of both the similarities and the differences found in systems is needed.

It is an exciting time in the area of computer systems. New technologies mean that architectures that were at one time interesting but not feasible are now feasible. Better software engineering means that we can consider several software alternatives, instead of 'more of the same old thing,' in terms of operating systems and system software. Faster and cheaper communications mean that inter-component distances are less important. We hope that this series contributes to this excitement in the area of computer systems by chronicling past achievements and publicizing new concepts. The format allows publication of lengthy presentations that are of interest to a select readership.

HERB SCHWETMAN

Foreword

To understand the problems raised by distributed working is to understand how tomorrow's information system will work. Just as there has come to be a standard structure for a car, agreed to by all manufacturers and well understood by all users, there is no doubt that these future systems will consist of machines of a wide variety of types distributed around a ring, ranging from individual work stations to a centralized file store, with general or special-purpose processors in between.

An information system can be visualized as a large factory for which a means has to be found for distributing and controlling the different tasks. Broadly speaking, this can be done by adopting either of two main principles. The first is to distribute the special tasks that have to be co-ordinated among different individuals or different teams; in this case the coupling between the tasks is fairly loose and the efficiency of the whole depends as much on the relative independence of the individual worker as on good co-ordination. The second approach is to break each task into a number of identical sub-tasks, in which case the coupling is stronger and it is important that these sub-tasks are well synchronised, even if a small amount of drift in the timing is allowed.

Taking two particularly well chosen examples to illustrate co-operation between different entities and control of identical tasks, respectively, Michel Raynal gives an excellent introduction to this type of problem in the first part, Chapters 1 and 2, of his book.

If we pursue the analogy with an industrial enterprise we soon see that the taking of decisions, especially when the fabric of the enterprise is woven from many units spread over a wide area, necessitates taking uncertainties into account. There will be spatial uncertainty, because there will never, at any given instant, be complete and exact knowledge of the location of all the constituent units; and temporal uncertainty because the information giving the conditions at, say, some subsidiary may be out of date by the time it reaches the decision-making centre. Thus if we try to imagine the processes that have to be implemented in order to mobilise, co-ordinate and stop the many activities of a distributed industrial enterprise we shall begin to understand some of the problems of distributed systems.

The aim of the second part (Chapters 3 and 4) of the book is to analyse the problems that arise in seeking to gain a knowledge of the topology of a distributed ring system and of the states of the entities that compose the system. The algorithms studied by Raynal, and here made readily accessible to the reader, are among the most recent and most important in this field. But the question arises, how can the head of the industrial enterprise put into operation the processes devised for the distribution and control of its tasks? If he has had the good fortune to have devised a way of working when the enterprise was small and all on one site – when he really could be everywhere and know everything that was happening – he may very well adapt this, with changes to suit the greater size and spread. His other course – which indeed is not incompatible with the first – is to make a careful study of the known and proven methods. This double approach is followed in the third part of the book: Chapter 5 is devoted to methods for distributing a constraint defined over the state of the system, assumed to be known instantaneously; whilst Chapter 6 gives a 'toolbox' of methods and mechanisms assembled by the author from a study of the already very voluminous literature.

As will already have been guessed from a reading of this analogy, the book says nothing about real, physical networks, machines or systems – there is no discussion of frames, of X-25 nor of 'Test and Set'. On the contrary, it is a study of, and reflection on, the general concepts, methods and mechanisms that are crucial in the design and implementation of distributed systems. The author has studied, and notably well assimilated, the considerable literature of the subject. Students will find here first class teaching material, and specialists a synthesis of the topics and an Ariadne's thread to guide them through the labyrinth of the many papers that will certainly be published on the subject in the very near future.

JEAN-PIERRE VERJUS

Preface

The field of distributed applications is constantly growing. This increase in the use of computer science as a preferred tool in ever more diverse areas is essentially the result of developments in both theoretical and practical aspects of the discipline.

The control of distributed applications and of the tools by which they are constructed is based on an understanding of the fundamental elements of what is known as the 'distributed system'. Compared to traditional systems (generally called 'centralized operating systems'), distributed systems differ in an essential way: the entities that form them cooperate in the achievement of a common aim by exchanging messages (there is no central memory to act as a locus for exchange), and, as a result, there is no global state in the system that can be detected instantly by one of these entities. To some extent, this can be seen as a relativistic perception of computer systems and data-processing applications.

This characteristic, combined with the development of hardware technology (local microprocessor networks) and with that of design methodology and software writing, shows that the control of distributed systems and of the applications developed on machine networks is based on a knowledge of concepts, tools, algorithms and specific methods.

The main object of this work is to introduce the elements fundamental to this knowledge. Each of its three parts tackles a particular aspect. The first, using two examples taken from the field of networks and applications (a data transfer protocol and the mutual dependence of logical clocks), concerns the problems caused by distribution and the tools required to solve them. The second part introduces algorithmic elements vital to any designer or user of distributed applications: network routes, learning distributed information, and the determination of the global state (this last point concerns the most basic problem posed by distribution). The third part considers the distribution of expressions and synchronization constraints with various hypotheses of reliability, and the contribution of *a priori* knowledge (such as the maximum delay in the transfer of messages, or the topological structure of the network) to the construction of applications and specific distributed systems.

The level of this book corresponds to that of postgraduate and advanced undergraduate students in computer science. In a more general way, it is intended for all those concerned with applications and distributed systems, networks and distributed computation. It is assumed that the general principles of centralized operating system design and the general theory of networks are understood. The algorithms described are self-explanatory (in a Pascal-like language for the sequential aspects).

The points made in this work are independent of the form or function of any specific distributed system. As indicated by the title, it is primarily concerned with the concepts, tools and algorithms that make up the foundations of distributed applications and systems.

MICHEL RAYNAL
Rennes (IRISA)—Lausanne (EPFL)
April–May 1986

Part 1

Introductory examples

This first part constitutes an informal introduction to the problems posed by distributed systems and networks. Two simple examples of distributed problems are described and studied. The nature of these problems enables us to attack the basic characteristics of partition systems and their solution enables us to introduce a number of algorithmic tools appropriate to the purpose.

These illustrations will help to gain a better understanding of the problems, algorithms, tools and methods that are studied in the later parts of the book.

First example: a data transfer protocol

1.1. The problem

The producer–consumer problem is a paradigm of system kernels and protocols encountered in networks. The problem can be stated very simply: an entity, called the producer, sends messages to another entity, called the consumer. The messages produced but not as yet consumed must be stored until their effective consumption; once transmitted, a message is, of course, consumed only once. For reasons connected with flow control between the producer and consumer, or with limited storage capacity for messages not yet consumed, it will be assumed that the maximum gap between the producer and consumer is of n messages.

1.1.1. Specification of constraints

Two operations (or procedures) are available—one to the producer and one to the consumer—for the sending and receiving of messages: *produce* and *consume*. The execution of these procedures must be synchronized so that a message produced will be consumed once and only once. To specify these constraints there are several possible tools; of these, counters have the advantage of simplicity. Consider the following two, associated with any procedure p:

$\#init(p) =$ number of executions of procedure p initiated from the start;
$\#term(p) =$ number of executions of procedure p terminated from the start.

Note that for these counters (initialized at 0):

i) they increase monotonically;
ii) $\forall p : \#init(p) \geq \#term(p)$
iii) their difference $\#init(p) - \#term(p)$ gives the number of executions of the procedure p currently in progress.

With these two counters it is easy to specify the constraints to which the procedures *produce* and *consume* are subjected.

In fact, $\#term(produce)$ gives the number of messages emitted by the producer and $\#term(consume)$ gives the number of messages absorbed by the consumer.

Since no transmission is possible if the producer is n messages ahead, and consumption is possible only if there are messages produced but not yet received, the conditions associated with the procedures are:

$$condition\ (produce) \equiv \#init(produce) < \#term(consume) + n$$
$$condition\ (consume) \equiv \#init(consume) < \#term(produce)$$

which allows the more 'canonical' expression:

$$condition(produce) \equiv \#init(produce) - \#term(consume) < n$$
$$condition\ (consume) \equiv \#init(consume) - \#term(produce) < 0$$

1.1.2. Problem invariant

When a call to the procedure p is authorized, the counter $\#init(p)$ is automatically increased by 1. As for each of the procedures *produce* or *consume*, the counter $\#init$ is given a positive coefficient, and in their canonical form the conditions are:

$$\#init(p) + x < y$$

and according to the assignment axiom:

$$(\#init(p) + x < y)\ \#init(p) := \#init(p) + 1\ (\#init(p) + x \leq y)$$

which allows the invariant to be deduced from:

$$\#init(produce) - \#term(consume) \leq n$$

and

$$\#init(consume) - \#term(produce) \leq 0$$

which, with the relationship:

$$\forall p : \#init(p) \geq \#term(p)$$

allows the invariant associated with the problem to be obtained:

$$0 \leq \#term(produce) - \#init(consume) \leq$$
$$\#init(produce) - \#term(consume) \leq n$$

If the events at the start and end of procedure execution are not of interest, it is possible to observe at another level, considering their executions as atomic, which is the same as making no distinction between the counters $\#init$ and $\#term$ by counting only the number of times that a procedure p has been executed. The counter $\#exec(p)$ is introduced for this purpose, and at any moment at this level of observation:

$$\#init(p) = \#exec(p) = \#term(p)$$

which gives the invariant:

$$0 \leq \#exec(produce) - \#exec(consume) \leq n$$

simply expressing the fact that the number of messages transmitted and not received varies between 0 and n.

1.2. A centralized implementation

Using the above specification, a number of applications are possible in a centralized context (this is a context based on the use of a shared memory). These implementations are distinguished by the tools on which they are based: semaphore, critical regions, monitors, etc. In this instance, the monitor has been chosen because, apart from the fact that this synchronization tool is available in several languages, it also constitutes a structuring tool for the synchronization kernels.

Storing messages

The storage of messages is carried out with a buffer of capacity n. Two pointers initialized at 0 allow this buffer to be managed: one is associated with the depositing of messages in the buffer, the other with their withdrawal.

var buffer : *array* $[0 \ .. \ n-1]$ *of messages*;
    in, out : $0 \ .. \ n-1$ *initialized to* 0;

In and *out* therefore index respectively the next empty and full locations in the buffer.

Implementation of control

The four counters used in the specification of the constraints to which the two procedures are subject are implemented by the four variables:

var startprod, endprod,
    startcons, endcons : $0 \ ... + \infty$ *initialized to* 0;

The control necessary to the correct execution of the two procedures p is expressed according to the following plan:

i) wait for the condition associated with p
ii) update the control variable *startp*
iii) act on the buffer (produce or consume)
iv) update the control variable *endp*.

Two queues equipped with *wait* and *signal* operators are used by the blocked processes as a waiting area when the preconditions for the continuation of execution are not satisfied.

monitor prodcons;
    start
            var buffer : *array* $[0 \ .. \ n-1]$ *of messages*;
                in, out : $0 \ .. \ n-1$ *initialized to* 0;

```
        startprod, endprod,
        startcons, endcons : 0 .. ∞ initialized to 0;
        queue p, c;
    procedure produce (m : message);
        begin
            if startprod − endcons ≥ n then p. wait end if;
            startprod : = startprod + 1;
            buffer [in] : = m;
            in : = (in + 1) mod n;
            endprod : = endprod + 1;
            if startcons − endprod < 0 then c. signal end if;
        end
    procedure consume (m : message);
        begin
            if startcons − endprod ≥ 0 then c. wait end if;
            startcons : = startcons + 1;
            m : = buffer [out];
            out : = (out + 1) mod n;
            endcons : = endcons + 1;
            if startprod − endcons < n then p. signal end if;
        end
end
```

Notes

This application, obtained systematically from the constraints, can be simplified. The pointers *in* and *out* are respectively *endprod mod n* and *endcons mod n*.

The unbounded character of the counters can be eliminated by introducing two integer variables:

var nbempty, nbfull : 0 .. *n initialized to n and* 0;

which have as their respective values:

nbfull = *endprod* − *startcons*
nbempty = *n* + *endcons* − *startprod*

and the appropriate control structure. This gives the 'classic' solution. It is nonetheless interesting to note that the first of the two centralized solutions obtained places the emphasis on the control of the procedures *produce* and *consume* whereas the second places it on the control of the resource constituted by the buffer. The reader is referred to works on centralized systems for further details on the various formulations.

1.3. A distributed application

1.3.1. Assumptions and structure

In a distributed context, the producer and consumer can no longer communicate messages via a central memory to which they will have mutually exclusive access. The only way they have of transmitting information between them is to use message transmission and reception primitives which operate on communication channels connecting the sites of the producer and consumer respectively.

At this level it is not important to know whether the channels are physical or logical; only the characteristics of their behaviour are of interest. In the context of this example, the following assumptions are made:

i) the producer and consumer are situated at two distinct sites connected by two unidirectional communication channels (or a single bidirectional channel);

ii) the channels are reliable. In other words, the behavioural properties of the links are as follows:

—no alteration of the messages;

—no duplication of the messages;

—no loss of messages (non-zero and finite transmission time);

no desequencing of messages (the messages are received at a site in the order in which they were transmitted from the other site, known as fifo – first in first out).

1.3.2. Events: their location and observation

The specification of constraints defining the problem shows that the condition to which the producer is subject allows it to produce up to n messages in advance (that is, not yet received by the consumer). To store these messages, a buffer of capacity n is introduced. Depending on the design chosen, this can be placed at the site of production, at the site of consumption (Figure 1), or at a site specific to it (Figure 2); or it can be distributed over various sites (to a maximum of n sites of capacity 1).

Note that, whatever the design chosen, on the abstract level there are only two sites, production and consumption; the arrangement is the domain of the distributed system designer.

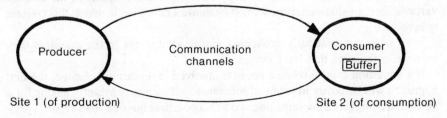

Fig. 1. A two-site implementation.

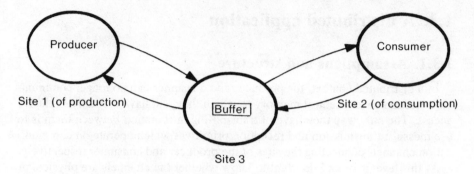

Fig. 2. A three-site implementation.

The *produce* procedure is used by the producer and, as a result the condition associated with it, must be evaluated at site 1. This condition involves the two counters #*init*(*produce*) and #*end*(*consume*). Consider the instance shown in Figure 1. The event which constitutes the start of production and to which the update of counter #*init*(*produce*) corresponds (increase of 1), is located at site 1. The end of consumption event, to which the updating of #*term*(*consume*) corresponds, is located at site 2. The sites must therefore exchange control information to ensure that the system functions correctly, that is that the constraints on production and consumption are satisfied.

The site at which an event occurs (start or end of an operation), and therefore that at which it is taken into account (by the modification of the counter which keeps track of the number of occurrences) is here distinct from the site at which the value of the corresponding counter is needed for the evaluation of a condition.

This distinction between the site at which an event takes place and where the observation of a variable gives its exact value and the site at which the value of this variable must be used is one of the basic characteristics of distributed algorithms and systems.

One solution is to provide each site that uses a variable, the modifications of which it does not manage (in other words, the events that bring about the modification of this variable do not occur on this site), with a copy of this variable, and to introduce a rule or protocol ensuring that the copy *changes* with the variable. In general, two approaches are possible:

— a site that manages a counter copy asks the site that manages the counter variable for the value each time it must evaluate a condition in which this counter is involved;

— all sites systematically broadcast any modifications to the counters they manage to the sites that have copies.

If a condition is false when a counter involved in it does not *change*, the first approach will be costly in terms of messages, with a site constantly asking for a value and being given the same one. To avoid pointless queries, the second is preferable: any modification of a system state variable is systematically broadcast to the parts of the system that need to know about it and manage copies. When a condition is false, the site that evaluates it waits to receive new values before evaluating it again.

Site 1 associated with production will therefore manage two counters:

$\#init(produce)$

and

$\#cterm(consume)$,

the latter being a delayed updated copy of the counter $\#term(consume)$ managed by site 2.

Site 1 can therefore evaluate at any time the following concrete condition:

$condition(produce) \equiv \#init(produce) - \#cterm(consume) < n$

Is this evaluation correct?

Since the counters are monotonic increasing, then:

$\forall\, p : procedure,\ x : counter : \#cx(p) \leq \#x(p)$

In addition, a difference between a counter $\#x(p)$ and a copy $\#cx(p)$ (which none of these sites can globally and instantaneously observe) shows that one or more increase messages for the copy are in transit between the site that manages $\#x(p)$ and that which manages $\#cx(p)$.

From the predicates:

$\#init(produce) - \#cterm(consume) < n$

and

$\#cterm(consume) \leq \#term(consume)$

it is possible to deduce:

$\#init(produce) - \#end(consume) < n$

or, in other words, each time site 1 evaluates the concrete condition of production in which a copy of the counter is involved, and finds it true, then the abstract condition is also true. The system is therefore functioning correctly as far as production is concerned and the same is true for consumption in the case of Figures 1 and 2 (with adequate copies).

This example illustrates one of the fundamental characteristics of a distributed system: the absence of a global state that could be instantaneously detected by one of the sites. On the one hand each site manages only those state variables that concern it, and on the other the delays in the transmission of messages between sites prevent a site from knowing in advance if the value of a copy it manages is the correct value of the corresponding variable or only one of its previous values. In the case where the variables are counters, the sites always have *a priori* knowledge: the behavioural characteristic of the variable (monotonic increasing), which here allows us to replace an abstract condition with a stronger concrete condition. (It will be seen in Chapter 5 that these results can be generalized to the case of variables whose the behaviour is arbitrary.)

1.3.3. Example of an implementation

The plan for distributed application proposed here corresponds to the structure described in Figure 1, in which one site is associated with each production or consumption entity; the buffer of capacity *n* intended to store messages transmitted but not yet received is entirely placed at the site of consumption (as an exercise, the reader can give the applications corresponding to other choices relative to the location of the buffer: at a third site, or on several other sites).

A given site—let it be the production site in this case—plays two distinct roles:

—firstly, that of production proper: if the site can produce, it sends a message (and updates the corresponding state variable, the counter #*init*(*produce*));

—secondly, the updating of the state variable #*cterm*(*consume*) when it receives a control message relative to the delayed increase of this copy of the counter.

These two roles played by a single site are those found in the network protocols, and which have been distinguished there in terms of layers in a reference model (the ISO model). This concerns the role linked to the service rendered by the site; transmission of messages produced, and the role of the management of control variables which requires the application of the service, and is linked to the system (Figure 3). These two activities of a site are universal and are found in distributed or centralized parallel systems and implementations in which there is an attempt at separating the control and processing parts. All this is simply an extension to the distributed context of the methodology of design by successive levels of abstraction, encountered in centralized programming, whether sequential or parallel.

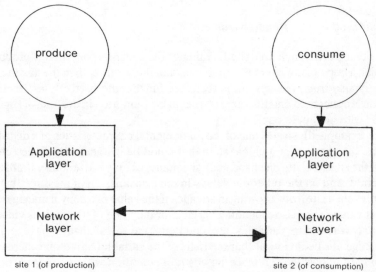

Fig. 3: Two-layered implementation.

The program texts of the sites are given in event form: the name of the event is given, with the condition under which it can occur and the sequence of calculations to which it leads (effect produced). These sequences are non-interruptible. The primitive *send* allows a message to be transmitted.

Production site

var startprod, ctermcons : $0 .. + \infty$ *initialized to* 0;
on a call to produce (*m* : *message*);
    possible only if startprod − ctermcons $<$ *n*;
    do
       startprod : = *startprod* + 1;
       send (*m*);
    end do
on reception of increase
    do
       ctermcons : = *ctermcons* + 1
    end do

The message *increase* received by the production site is a pure signal (non-valued).

Consumption site

At this site the only state variables retained are the counters actually used, namely *startcons* and *endprod*, which have a well defined value at this site. There is no point in managing the counter *endcons*, which is not really involved in the consumption condition; on the other hand, its copy at the other site *is* used. The sending of the corresponding control messages *increase* must therefore be managed.

var startcons, termprod : $0 .. + \infty$ *initialized to* 0;
    in, out : $0 .. n-1$ *initialized to* 0;
    buffer : *array* $[0 .. n-1]$ *of messages*;
on a call to consume (*m* : *message*);
    possibly only if startcons − termprod $<$ 0;
    do
       startcons : = *startcons* + 1;
       m : = *buffer* [*out*];
       out : = (*out* + 1) *mod n*;
       send (*increase*);
    end do
on reception of message m
    do
       buffer [*in*] : = *m*;
       in : = (*in*+1) *mod n*;
       termprod : = *termprod* + 1;
    end do

Notes

As with the centralized version, the program texts can be simplified either by using the counters *startcons* and *termprod*, taken **modulo** *n*, as input and output pointers into the buffer, or by introducing a state variable that sums the two counters:

var spaceoccupied = *termprod* − *startcons*

The reader may find it of interest to compare the two solutions presented. Such a comparison will show the basic differences between centralized and distributed systems: the latter are marked on the one hand by an approximate knowledge of certain state variables and the absence of a global state, owing to the absence of central memory (which would define a global state that could be detected instantaneously by mutually exclusive memory access); and on the other by the use of messages transmitted in a non-zero finite time.

The assumptions on the behaviour of the channels will now be modified, and the case of a non-reliable environment will be examined.

1.4. The unreliable environment case

1.4.1. Assumptions concerning channels

In the previous distributed solution of the producer–consumer paradigm, it was assumed that the communication channels were reliable. This assumption is now modified in that the channels display the following behaviour: they may
— lose messages;
— duplicate messages;
— desequence messages;
— alter messages.

The aim is to achieve a distributed algorithm or protocol which fulfils the same function as before; that is, at any moment:

$$m_1 \; m_2 \; m_3 \; m_4 \ldots m_i \; m_{i+1} \ldots m_j \ldots m_k$$

is the sequence of messages that have been produced, and the sequence of messages consumed is:

$$m_1 \; m_2 \; m_3 \; m_4 \ldots m_i \; m_{i+1} \ldots m_j$$

with $k - n \leq j \leq k$ (there is a maximum of *n* messages produced and not yet consumed).

The existence of an altered message detection mechanism is assumed. Such detection leads to the rejection of the message received; by this device the problem of alteration is reduced to that of loss. In general, therefore, if the problem of loss can be resolved and if altered messages can be detected, the problem of message alteration can be solved.

It is assumed, in addition, that the two channels between the sites of production and consumption are equivalent; this means that if a single message is constantly sent backwards and forwards along the channel, it will reach its destination at least once. The result of this hypothesis is that a channel cannot lose all the messages it transports; the hypothesis does not, however, include the case of a link that loses all its messages because of being broken.

1.4.2. Fundamentals of the solution

1.4.2.1. SEQUENCE NUMBERS

To resist loss, desequencing and duplication of messages, any solution must have two aspects: the messages themselves and the processing to which they will be subjected.

The site of production manages a counter, initialized to 0 and increasing monotonically, which associates a sequence number with each of the messages transmitted. A message is formed by the pair:

(sequence number, message value)

The first message sent will have the number 1, the second 2, and so on. Duplication is easily detected: it is indicated by the reception of messages with the same sequence number. The same applies to the detection of desequencing, when the message sequence numbers are not presented in increasing order.

1.4.2.2. BASIC PROTOCOL

The introduction of sequence numbers is not enough to solve the problem entirely. Consider the case of message loss: at a given moment a gap in the numbering relative to the sequence of received messages can indicate either that the corresponding message has been lost, or that it has been overtaken by messages with higher numbers and will arrive at its destination later. The tools so far introduced are therefore not sufficient except in cases where the messages may be duplicated and desequenced, but not lost.

To solve the problem presented by loss, a protocol is introduced. This consists of rules and the transfer of supplementary control information between the communicating entities. The receiving entity, which is here the site of consumption, will transmit acknowledgements of receipt of messages. Since an acknowledgement relates to a given message, the acknowledgements will be identified by the sequence number of the corresponding message. This is a protocol of PAR (positive acknowledgement retransmission) type; that is, one based on the fact that the receiver does not transmit a negative acknowledgement of reception ('non-reception' acknowledgement); only acknowledgements of reception relative to the messages received are transmitted (these are called positive acknowledgements).

One solution consists of the producer entity systematically retransmitting the messages that have not been acknowledged after a certain time following their initial transmission. This involves a guard clock system (timeout) that must be managed by the transmitter.

It is important to note that these operating rules, or protocols, may introduce message duplications. In fact, depending on the calibration of delay before retransmission, a message may be retransmitted (one or more times) while the acknowledgement relative to its first transmission is in transit or even before the original message has reached the receiver site. The solution proposed here to this problem of loss assumes, therefore, that the duplication has already been solved. (It is of great importance that the solutions should not depend circularly on each other.)

The same assumptions are made concerning the channel that carries the messages and the one that carries the acknowledgements. These may also be duplicated, desequenced and lost. The same problem arises again, and if it were necessary, in order to solve it, to introduce acknowledgements of the acknowledgements and so on, then there would be an infinite recurrence of the problem, which would have no solution.

Consider the problem of duplication: an acknowledgement of reception received several times does not present a problem, if only the first occurrence is considered. As for desequencing and loss, the above protocol ensures that if the producer entity has not received the expected acknowledgement after a certain period of time, it will retransmit the corresponding message. The loss of an acknowledgement is therefore assimilated by the protocol to the loss of the corresponding message. The principle of the solution is therefore sound.

Before introducing other tools, the problems encountered and the tools introduced to solve them can be summarized:
— duplication, desequencing : sequencing numbers;
— loss : acknowledgements of reception, guard clocks.

1.4.2.3. CALIBRATION OF THE GUARD DELAY

It is important to note the role of delay for the guard clock. As far as the logic of the solution is concerned (that is for its correction) the delay is necessary, but it can be calibrated in any way between $]0, +\infty[$. In operation, however, the same does not hold true; too small a calibration would lead to constant retransmission and therefore to many duplications, thus loading the channels unproductively. In the same way, too large a calibration would err in the other direction: there would be no duplication, but a lost message would be retransmitted after a long delay and the channels would be underused. For reasons opposite to those given above, the result would be the same: poor performance by the system.

The calibration of the guard system is therefore a crucial point in obtaining a high-performance system: the determining parameters are the mean and variance of the message transfer time.

1.4.2.4. WINDOWS AND CREDITS

The messages – structured as ordered pairs (sequence number, message value)—which are transmitted but not yet acknowledged must be retained by the transmitter since it may have to retransmit them at the end of the fixed delay. Generally speaking, it is assumed that the transmitter can transmit a maximum of ct messages with consecutive numbers before receiving the corresponding acknowledgements; this number is called the *credit* associated with the transmitter.

Consider the following arrangement of sequence numbers from the point of view of the transmitter; these numbers relate to the following messages:

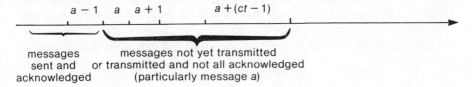

The numbering interval $[a, a+(ct-1)]$ is called the *window* of the transmitter; a is its origin and ct the width. When the acknowledgement of message a arrives at the transmitter, the origin of the window moves to b, the number of the next unacknowledged message; the window then becomes $[b, b+(ct-1)]$.

Similar concepts apply to the receiver: it has a credit cr and a window $[r, r+(cr-1)]$, of origin r and length cr, associated with the following structure, these sequence numbers denote:

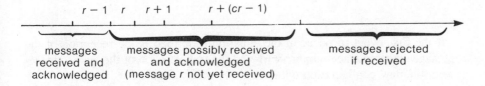

When message r has been received by the transmitter, it acknowledges it and the reception window moves to $[s, s+(cr-1)]$ where s indicates the number of the first message not yet received.

The windows are such that the relationship $a \leq r$ is invariant; they evolve with the arrival of messages and acknowledgements. The credits allow the two entities of transmission and reception to be controlled; the transmission credit ct fixes the maximum number of different messages in transit and means that the transmitter is controlled by the receiver; the credit cr specifies the size of the memory provided by the receiver to store the received messages. These parameters were introduced in order to achieve what is known as flow control in networks. There are various types of strategy for this type of control. Depending on the values of the credits, two cases should be considered:

— $ct = cr$: this is called a worst-case strategy;
— $cr < ct$: this is called an optimistic strategy.

In the latter case, the receiver may disregard messages with a sequence number not included in its window; it is to be hoped that this situation will not arise frequently and that the channel will have the capacity to store $ct - cr$ messages.

As an example, consider the following situation (Figures 4 and 5). The transmitter credit, ct, is 5 while that of the receiver, cr, is 2. The windows of the transmitter and receiver are respectively $[x, x+4]$ and $[x+1, (x+1)+1]$. The transmitter has transmitted messages up to number $x+4$ and has received acknowledgements of messages up to number $x-1$; the receiver has received and acknowledged all messages up to number x (inclusive), has received and acknowledged that of number $x+2$, and received and discarded that of number $x+4$ (Figure 4).

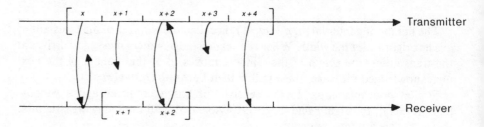

Fig. 4. Situation before reception of message number $x + 1$

If the first event to occur in the situation described is the reception of the message with sequence number $x+1$ and the transmission of the acknowledgement, the new configuration obtained is shown by Figure 5.

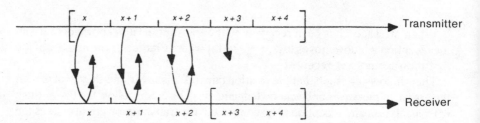

Figure 5. Situation after reception of message number $x + 1$.

Reception of the acknowledgements x and $x+1$ will cause the movement of the transmitter window to $[x+3, (x+3) + 4]$.

The introduction of the credit and window concepts, jointed to that of sequence numbers, will allow a solution to be given for the problem under consideration.

1.4.3. Structure of the solution

1.4.3.1. CHOICE OF IMPLEMENTATION

In the previous section, the bases for solution of the problem were introduced: sequence numbers, the PAR protocol, and credits.

In order to reduce the number of messages transmitted in the case where acknowledgements are lost, the previous protocol can be better adapted to the problem by modifying the semantics of the acknowledgement. An acknowledgement carrying the sequence number x of the message whose reception it acknowledges will not be transmitted unless all the messages with numbers less than x have been received and acknowledged. As a result, when the transmitter receives the acknowledgement numbered x it knows (without additional information having to be given), that the messages with lower numbers have been received, even if it has not received the corresponding acknowledgements.

To simplify the implementation, instances of desequencing will be treated as errors: as a result the receiver will not accept a message unless it has received all the messages with lower numbers; in other words $cr = 1$. The credit associated with the transmitter is ct. In the case where $ct = cr = 1$, the solution obtained is similar to the alternating bit protocol.

1.4.3.2. STRUCTURE

The system is made up of two sites (Figure 6). The site of production is equipped with a buffer of capacity n in which the messages are deposited before being transmitted ($n \geq ct$). This allows the producer (which calls the *produce* procedure) to be desynchronized from the transfer protocol; this is the distinction made in Section 1.3.3.

The site of consumption is structured in the same way; a queue of capacity n allows the messages received to be stored.

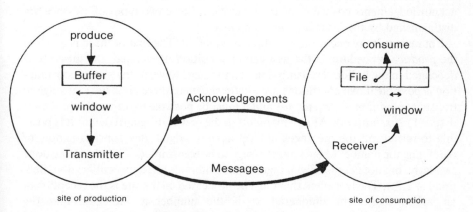

Fig. 6. Structure of the system.

1.4.4. An implementation

1.4.4.1. THE SITE OF PRODUCTION

The site of production is provided with the following variables, initialized to 0:

var nextseqno, sendnum, acknum : 0 .. + ∞ *initialized to* 0;
   full : 0 .. n *initialized to* 0;
   buffer : **array** [0 .. *n*−1] *of messages*;

the variable *full* indicates the space occupied in the buffer. It evaluates to the expression: #*init*(*produce*) − #*cend*(*consume*) used in the condition associated with production.

The three other integer variables are counters which define respectively:

—*nextseqno*: the sequence number of the next message received in the buffer via a *produce* operation;

—*acknum*: the smallest awaited acknowledgement number;

—*sendnum*: the number of the next message to be sent to the site of consumption. The following must always apply:

$$sendnum \in [acknum, \; acknum + (ct-1)]$$

The transmission window is therefore:

$$acknum \qquad sendnum \qquad\qquad acknum + (ct - 1)$$

The messages transmitted are ordered pairs, as stated above, but the acknowledgements consist of a single number. These two types of message are distinguished by the indicators *mes* and *ack*.

Four events may occur on the site of production. The first of these is a call to the *produce* procedure; if the associated condition is satisfied, the message is deposited in the buffer. Transmission of a message towards the site of consumption is possible if two conditions are jointly satisfied: there must be a message to transmit (*sendnum* < *nextseqno*) and it must be possible to transmit it (*acknum* + (*ct*−1) ≥ *sendnum*). At any moment (at the end of the guard delay) it is possible to retransmit the messages not yet acknowledged: this consists simply of redefining the number of the next message to be transmitted. This should be done regularly, but not necessarily at regular time intervals (cf. delay calibration, Section 1.4.2.3.). The last event that must be processed on the site is the reception of an acknowledgement numbered *x*; if the number *x* indicates that the acknowledgement has not been overtaken by others, it is taken into account and the full variable is updated accordingly by reducing it by $x+1 - acknum$. (In the following procedure the symbol $- -$ indicates comments.)

> **on a call to produce** (*m* : **message**)
>> **possible only if** *full* < *n;*
>>> **do** *full* : = *full* + 1 ; $- -$ #*init*(*produce*) : = #*init*(*produce*)+1;
>>> *buffer* [*nextseqno* **mod** *n*] : = *m*;
>>> *nextseqno* : = *nextseqno* + 1;
>> **end do**
>
> **on transmission**
>> **possible only if** (*sendnum* < *nextseqno*) *and* (*sendnum* ≤ *acknum*+(*ct*−1));
>>> **do**
>>> **send** *mes*(*sendnum, buffer* [*sendnum* **mod** *n*]);
>>> *sendnum* : = *sendnum* + 1;
>>> **end do**
>
> **on retransmission**
>> $- -$ *always possible and performed regularly*
>>> **do**
>>> *sendnum* : = *acknum*
>>> **end do**
>
> **on reception of** *ack*(*x*)
>> **do**
>>> **if** *x* ≥ *acknum*
>>>> **then** *full* : = *full* − (*x*+1 − *acknum*);
>>>> $- -$ #*cterm*(*consume*) *is incremented by the appropriate value*
>>>> *acknum* : = *x*+1
>>> **end if**
>> **end do**

1.4.4.2. THE SITE OF CONSUMPTION

The queue *f* placed at this site is equipped with two access operators *deposit* ((*x,y*),*f*) and *withdraw* (*f*). These operators allow an element to be added and removed respectively, according to a fifo discipline. The elements placed in the file are orders/pairs (sequence number, message value).

Note that the service rendered is that described at the start of this chapter, in Section 1.1.1: the number of messages produced cannot exceed by more than *n* the number of messages consumed. The control of the producer is expressed in the condition associated with the *produce* procedure; the following must apply for *produce* (cf. Section 1.3.2.):

$$\#init(produce) - \#cterm(consume) < n$$

which is expressed at the production site by *full* < *n*. Therefore *full* must be managed so as to satisfy:

$$full = \#init(produce) - \#cterm(consume)$$

for the solution to be correct. An increase of *full* corresponds to the start of a production and its decrease corresponds to the reception of a new acknowledgement number. At the site of consumption, therefore, the acknowledgement number x must be sent once the message numbered x has been dealt with and not before (otherwise the gap between producer and consumer could be greater than n). Once transmitted, an acknowledgement can be retransmitted to make up for its possible loss (this behaviour is not, however, necessary to the correct functioning of the system, as seen in Section 1.4.2.2.).

In addition, two other types of behaviour are possible at the site of consumption. The first corresponds to the event constituted by a call to the *consume* procedure, which can take place only if queue f is not empty. The second is associated with the reception of a message from the site of production: if the sequence number is the one expected, the message is stored in the queue; if not it is discarded. (Note that the credit *cr* is equal to 1.)

The context of the site is the following:

var f : *queue*(*n*) *initialized to empty*;
  nbpres : $0 .. n$ *initialized to 0*;
  noexpected, noack : $-1 .. + \infty$ *initialized to* $0, -1$

The variable *nbpres* gives the size of *f*, that is the number of messages locally available for consumption; it has the value:

$$\# term(produce) - \# init(consume)$$

used in the condition of consumption; the variable *noexpected* designates the sequence number of the next message expected, and *noack* is the number of the last acknowledgement sent (and to be returned) to the site of production (this is the value which, on arrival at the site, will represent $\# cterm(consume)$).

on a call to consume (*m* : *message*)
    possible only if nbpres < 0;
    do
        nbpres : = *nbpres*-1;
        $--$ $\# init(consume)$: $= \# init(consume)+1$;
        (*noack, m*) : $=$ *withdraw*(*f*);
        send ack (*noack*);
        $--$ *send* ($\# term(consume)$)
    end do
on reception of mes (*seqno, mesval*)
    do
        if seqno $=$ *noexpected*
        then nbpres : $=$ *nbpres*$+1$;
            $--$ $\# term(produce)$: $= \# term(produce)+1$;
            deposit((*seqno, mesval*), *f*);
            noexpected : $=$ *noexpected*$+1$
    end if

```
      end do
on retransmission
   – – always possible and performed regularly
      do
            send ack (noack);
      end do
```

1.4.4.3 FLOW CONTROL

If the aim is to control the load on the communication channel rather than tie the consumer to the producer, the above solution must be modified, because then the implementation of the service performed is constrained by a maximum value of n and not by the service itself.

The site of production can remain as it is, but the site of consumption must have a queue whose capacity is *a priori* infinite, so that the consumer cannot block the producer. In addition, since there is no longer any need to indicate the increase of the counter #*end*(*consume*), the sending of acknowledgements can be carried out as soon as the corresponding message has been received.

The site of consumption, in which queue *f* contains only the message values and in which the variable *nbpres* varies between 0 and $+\infty$, behaves as follows:

```
on a call to consume (m : message)
   possible only if : nbpres > 0;
   do
      nbpres : = nbpres−1;
      m : = withdraw (f);
   end do
on reception of mes (seqno, mesval)
   do
      if seqno = noexpected
         then nbpres : = nbpres +1;
              deposit (mesval, f);
              noexpected : = noexpected+1;
              send ack (noexpected−1);
      end if
   end do
on retransmission
   do
      send ack (noexpected−1);
   end do
```

As an exercise, the reader may: i) compare the above solution with the almost identical solution in which only the first two modes of behaviour are considered— that is without the retransmission sequence; and ii) compare the solution obtained with the same solution in which the instruction:

send ack (*noexpected*−1)

is placed not in the *if . . . end if* loop but just after it.

The first of these two modifications leads to an incorrect solution: if the production buffer is full and the acknowledgement of the last message is lost, the protocol is blocked, with the transmitter retransmitting messages with numbers lower than the one being considered by the receiver. The modification brought about by the second solution corrects it, because the last acknowledgement is retransmitted after the reception of any message. The message and acknowledgement numbers are no longer the same as in the initial solution: it is very interesting to know in which circumstances each performs better. These protocols therefore allow reliable transmission service (producer–consumer structure) to be set up on unreliable channels.

1.5. Bounding the ranges of variation

1.5.1. Counters

1.5.1.1. FURTHER COUNTERS

The counters so far introduced keep a count of the number of occurrences of the events with which they are associated. Depending on the problem to be processed, it is possible to introduce new ones, which, when added to $\#init(p)$ and $\#end(p)$, allow more sophisticated synchronization constraints to be expressed. The counter $\#req(p)$, for example, which counts the number of calls to the procedure p made but not necessarily authorized from the point of view of their execution, allows certain problems of priority to be resolved (the expression $\#req(p) - \#init(p)$ gives the number of the call to p currently waiting).

1.5.1.2. MODULO IMPLEMENTATION

The major drawback with counters when they are used for the specification and application of systems is their potentially infinite growth. If each counter is considered individually, it is not possible to limit its range of variation; if, on the other hand, the expressions in which they are involved are examined, this is not necessarily so.

Consider the specification stated in Section 1.1.2. The invariant:

$$0 \leq \#term(produce) - \#init(consume) \leq$$
$$\#init(produce) - \#end(consume) \leq n$$

shows that each of the two expressions used in the conditions is formed of two counters, the difference between which is limited, and varying between 0 and n. In addition, the step by which each counter increases is 1. These two elements allow a *modulo k* implementation, by choosing a value of k greater than the maximum difference between the two counters and by redefining the new constraints.

Example

By way of illustrating the schema of the previous implementation with simpler notation, consider two counters *a* and *b*, and constraints on procedures *p* and *q* expressed as:

condition $(p) = a - b < n$
condition $(q) = 0 < a - b$

From these conditions, the invariant $0 \leq a - b \leq n$ can be derived (with the same form as the previous invariants). The counters are increased by steps of 1.

The set of values that the counters (a, b) can pass through while respecting the constraints can be represented by the region bounded by the lines $a-b = 0$ and $a-b = n$:

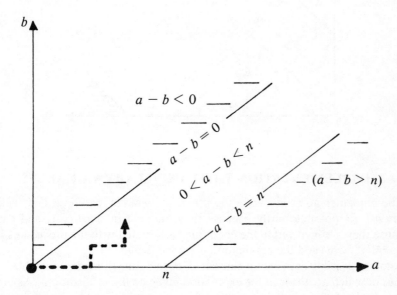

By taking $k = n+1$, the counters *a* and *b* can be put into action respectively by the variables:

var va, vb : $0 \ldots n$;

which take values *mod* k : $va = a$ *mod* k and $vb = b$ *mod* k.

The conditions associated with the procedures become:

condition $(p) = ((va - vb) \neq n)$
condition $(q) = (0 \neq (va - vb))$

The first can be rewritten taking account of the fact that $n = -1$ *mod* k:

condition $(p) = ((va - vb) \neq -1)$

The set of values that the pair (*va*, *vb*) can take is therefore defined by the following part of $(n \bmod k)^2$:

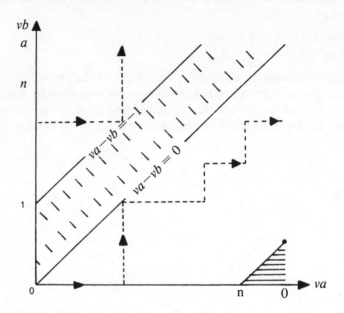

1.5.1.3. IMPLEMENTATION WITH ANCILLARY VARIABLES

The implementation involving the previous counters, based on their values *modulo k*, is possible partly because they increase in steps of 1, and partly because they are involved in the conditions associated with the procedures in a differential form (and the maximum difference defines *k*).

Another possible application consists of considering (bounded) variables, the values of which are those of the expression involved in the conditions. As in the centralized solution given in Section 1.2, two variables can be introduced:

var nbfull, nbempty : 0 .. *n initialized to* 0, *n*;

which must be managed so as to satisfy the following values:

nbfull = #*term*(*produce*) − #*init*(*consume*)
nbempty = *n* + #*term*(*consume*) − #*init*(*produce*)

and this gives:

condition(*produce*) = *nbempty* > 0
condition(*consume*) = *nbfull* > 0

The introduction of these variables does not altogether solve the problem. In fact, the variable *nbempty* placed at one of the two sites cannot give an exact value, since it counts events taking place on two distinct sites. The solution to this is to place at the site the variable *apnbempty*, which gives an approximate value of the abstract variable *nbempty*, such that:

$$apnbempty > 0 => nbempty > 0$$

It is placed at the site of production, which will not increment it until the site of consumption sends the authorizing messages (increase of #*end*(*consume*)). The variable *full*, handled by the site of production as in Section 1.4.4.1, is such a variable giving an approximate value, and is used in the production condition.

1.5.1.4. COMPARISON OF THE TWO IMPLEMENTATIONS

The two solutions suggested for limiting the domain of counter variation are based on the following assumptions:
—incrementation by steps of 1;
—bounded differential increases.

These are, in reality, two implementations of the same solution, the essence of which is to detect a difference in the counters. What differentiates them, therefore, relates only to the implementation.

In the first solution (*modulo k* counters), the evaluation on a given site of the condition associated with the controlled procedure can be made at the same time as the incrementation of the local counters involved; in other words, only access to the variables taken individually must be mutually exclusive, and the evaluation of an expression does not require overall exclusion of the set of variables involved. For example the counter #*cterm*(*consume*) (cf. Section 1.3.3.) can be increased by the site of production during evaluation of the condition:

$$\#init(produce) - \#cterm(consume) < n$$

The second solution does not offer the same advantage. The variable *apnbempty* is both increased and decreased by the site of production, consequently these two operations must be carried out in mutual exclusion (which, in the general case, justifies the hypothesis of atomicity of the processing carried out by a given site). For example, in Section 1.4.4.1, the variable *full* must be handled in exclusion on the site of production by the process associated with the production (incrementing by 1) and the process associated with the reception of acknowledgements (decrementing).

The two implementations are therefore of interest because of their different properties, owing to different representations of the differential counter incrementation: either explicit or contracted into a single variable. These two implementations correspond, in a distributed context, to the two solutions mentioned in Section 1.2 in a centralized context: the first (the counter) is oriented towards the process sites, the second (bounded variable) towards the resource constituted by the channel.

1.5.2. Sequence numbers

1.5.2.1. THE PROBLEM OF DUPLICATION

The algorithm introduced in Section 1.4 to deal with an unreliable environment (duplication, desequencing and message loss) is based on a PAR-type protocol and on the use of sequence numbers to create unique and ordered identities for the various messages. As with the previous counters, these numbers present the disadvantage of unbounded growth.

If these numbers are constrained to vary within the domain $0 \, . \, . \, m-1$ (that is, *modulo m*), a problem appears: that of avoiding the confusion of a message numbered x with a 'double' (arising from retransmission of a duplicate) numbered $x-m$ (or, more generally, $x-km$). Over the interval $0 \, . \, . \, m-1$ all these messages have the same effective identity.

It is important to note that this problem did not arise in the specification of the producer–consumer schema in which only the counters are used. It appears only in the implementation in which there is a need to deal with unreliable (but still unbiased) channels.

1.5.2.2. DEFINITION OF THE RANGE OF VARIATION

The first step is to define a range of variation for the numbers. The constraints that must be satisfied if this range is to be used safely will be examined next.

A general framework is considered in which, at transfer protocol level, the two entities are provided with the following credits and windows:

—transmitter: $[t, \, t+(ct-1)]$ (numbers of messages that can be transmitted)
—receiver: $[r, \, r+(cr-1)]$ (with $cr \leq ct$); (numbers of messages that will be accepted as received).

As a result of the way they are used, the two windows change so that they include (noting that $t \leq r$ and $ct \geq cr$):

—either some common numbers (of messages transmitted and not yet received):

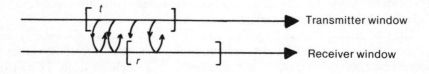

—or adjacent sequences of numbers without overlap (all messages transmitted have been received and acknowledged, but the acknowledgements have not yet reached the transmitter):

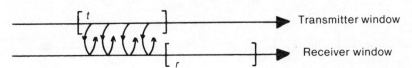

If the assumption (to be abandoned later) is made that there is no duplication, the maximal interval on the message numbers considered (sent or awaited) at a given moment is $ct + cr$ or $2ct$, taking the largest possible value for cr. Under these conditions, since t is the left-hand side of the transmitter window, the sequence numbers that may appear in the system at a given moment correspond to the interval $t \ldots t+2ct-1$, with width $2ct$. If the messages are numbered cyclically within the interval $0 \ldots m-1$, then if $m > 2ct$ the same number x will not appear several times, associated with different messages, in either of the windows with total width $2ct$.

If m defines the length of the numbering cycle $(0 \ldots m-1)$, the credit associated with the transmitter must always be less than $m/2$ for a solution to be possible. The assumption of non-duplication must be discarded if the problem is to be entirely solved.

1.5.2.3. TIMING CONSTRAINTS

The transfer algorithm is based on the systematic retransmission of messages the acknowledge of which has not yet been received, so duplication is an inevitable part of it. The aim is to obtain a mode of operation for the receiver in which it will accept any message, whether new or repeated, that has a number x lying in its window.

Let $x_1, x_2, x_3, \ldots$, be the order numbers of successive messages with the same sequence number x, with $x \in 0 \ldots n-1$. Then:

$$x = x_i \bmod m$$

and

$$x_{i+1} - x_i = m$$

Consider a case where the message with order number x_i is being transmitted, and has its sequence number x in the receiver window. If $m > 2ct$ all the 'old' messages with x numbers in the receiver window have been received and acknowledged, and the acknowledgements received by the transmitter. It this were not the case, then according to the rules defining transmitter behaviour, the window would not have been able to move forwards, and this is not the case here.

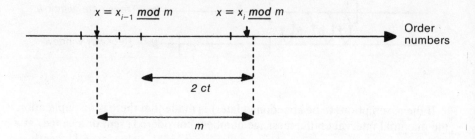

When a message numbered x is received, the 'old' messages of the same number have been received and acknowledged. Their possible duplicates, however, may still be in transit. Consider the following configuration:

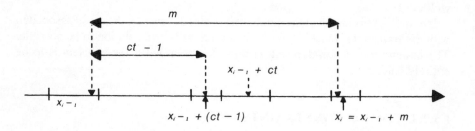

The transmitter sends out duplicates of the messages already sent and for which it has not received acknowledgement. The window is therefore:

$$[x_{i-1} \; mod \; m, \; (x_{i-1} + (ct-1)) \; mod \; m]$$

and moves to:

$$[(x_{i-1} +1) \; mod \; m, \; (x_{i-1} + ct) \; mod \; m]$$

on reception of the acknowledgement corresponding to the message with number $x(x_{i-1} \; mod \; m)$. Henceforth it will not send any more duplicates of this message; the last duplicate for this message will have been sent, if at all, at the latest just before the message with number $(x_{i-1} + ct) mod \; m$. To prevent any problems when the number $x(x_i \; mod \; m)$ again corresponds to the receiver window, these duplicates must no longer be in transit. This can be taken care of by adding a restriction on the maximum delay of transmission Δ for a message (a 'duplicate'). If the receiver accepts the messages in number order (which is the same as having $cr = 1$), Δ must be less than the time separating the allocation of the number $((x_{i-1} + ct) \; mod \; m)$ (the instant at which this number is allocated corresponds, in the worst case, to the departure of the last duplicate of number $x_{i-1} \; mod \; m$) from the allocation of number $y_1 = ((x_i -1) \; mod \; m)$. In the worst case, the message carrying this number y_1 is received and acknowledged immediately and the receiver window then includes the number x_i again at its right-hand end.

The distance between these two numbers is therefore $m-(ct+1)$. In the case where $cr = ct$ and where the receiver acknowledges the messages with numbers in the window, the worst case occurs if the windows are adjacent: the allocation of the number $y_2 = x_i - ct$ and the reception of the corresponding message would make x_i re-enter on the right-hand side of the receiver window. Since the last duplicate for number x_{i-1} would leave just before the allocation of number $x_{i-1} + ct$, the distance between these two numbers would be $m - 2ct$. In the general case, it is equal to $m-(ct+cr)$.

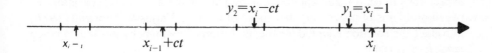

Note that the messages can be desequenced; that is, they can pass each other on the channel. The worst case would therefore arise if the duplicates travel as slowly as possible and are overtaken by other messages.

The maximum delay in message transmission must therefore be less than the time separating the allocation of two numbers placed a certain distance away from each other, if the problem posed by duplication is to be resolved. Added to the constraint defining the domain $(m>2ct)$, this means that only bounded variables are involved in this implementation.

The desequencing considered above can be of two possible types: overtaking by messages on the channels, and the introduction by the retransmission protocol of duplicated messages with 'out of sequence' numbers.

In the case where the only possible desequencing is as a result of retransmission, the above constraints can be redefined. The last duplicate of x_{i-1} will have left, in the worst case, between the message with number $x_{i-1} + (ct-1)$ and that with number $x_{i-1} + ct$. Since the channel does not desequence the messages, when the message number $x_{i-1} + ct$ arrives at the receiver, the number x can reappear in its window without there being any risk of confusion. The reader is recommended to find the associated constraint as an exercise, as well as the formulation for the corresponding protocols (the test $x \geq acknum$ carried out during reception of an acknowledgement in Section 1.4.4.1 is no longer necessary, for example, since the acknowledgement numbers received form a non-decreasing sequence).

1.6. References

The producer–consumer problem is a classic among the problems encountered in system design. Centralized solutions will be found in works on operating systems (Peterson & Silberscharz [1983], Krakowiak [1985]). The interest of this problem lies not only in its practical usefulness, but also in the simplicity of the way it can be stated and the various analyses to which it lends itself. It constitutes a paradigm for parallelism in general, and distribution in particular (André *et al.* [1983], Lamport [1984], Plouzeau [1987]).

The conceptual tool constituted by the counters associated with controlled procedures was introduced by Robert and Verjus in the context of centralized applications (Robert & Verjus [1977]). Its use was extended to the distributed context in Bochmann(1979); André(1983) introduced a methodology for its use in parallel programming, which provided the inspiration of Section 1.3 of this chapter. Raynal (1983) provided a comparative study of counters and path expressions in centralized applications by placing the accent on separation between problems of processing and those of control. Verjus & Thoraval (1986) introduce the solution adopted for the limitation of counters by introducing their *modulo n* variables.

The modular analysis of systems has made great progress based on the work of Dijkstra, Parnas and others, and the introduction of abstract types into programming languages. For networks, a similar effort of structuring was carried out for so-called 'open' systems and resulted in the definition of the ISO layered model (Zimmerman [1980]).

Sequence numbers and the PAR protocols were initially introduced into networks. The same is true of the window concept (Cerf & Kahn [1974]). The transfer algorithm, in the case of an unreliable environment considered above, is a variant of the distributed algorithm known as Stenning's protocol (Stenning [1977]). The protocol known as the alternate-bit protocol (Bartlett *et al.* [1969]) constitutes a simplified version in which the transmitter credit has a value of 1. Generally speaking, a message in the direction from A to B can be carried by an acknowledgement, travelling in that direction, of a message from B to A, which itself can carry the acknowledgement of a message travelling from A to B. The protocols obtained are therefore general and ensure duplex transfer.

On a less specific level, the reader will find information on networks in Tanenbaum (1981) and Pujolle *et al.* (1985), an introduction to distributed computation in Raynal (1985a) and techniques concerning distributed systems in Cornafion (1981).

Second example: mutual control of logical clocks

2.1. The problem

This chapter concerns a second problem which illustrates other aspects of distributed systems and networks. The problem of the producer, considered in the previous chapter, introduced certain algorithmic tools (counters, specific protocols based on acknowledgements, sequence numbers) and mentioned the incidence of non-reliable behaviour in the connections between the communicating entities. In this problem, the two entities behave in different ways. The producer transmits messages and the consumer absorbs them, with two program texts taking account of this. In the example treated in this chapter, all the entities will respect the same behaviour and only one text will, therefore, be necessary.

Consider two entities (or processes) P_i and P_j, *a priori* independent. Their only means of interaction is the exchange of messages. Each of them has a variable h_i called the logical clock, handled as a counter increasing by steps of 1. When a process P_i transmits a request (to a resource, or server etc) it adds an identifier formed by the ordered pair (logical date, place) $= (h_i, i)$. This pair will uniquely identify the request (as the h_ith request of process P_i) if, before transmitting a request, a process systematically increases its clock before taking the value.

The problem is as follows: the two processes are to be constrained such that at any moment neither of them can transmit more than δ requests ahead of the other. In view of the association of the clock values with the requests, the problem consists of keeping the following relationship invariant:

$$|h_i - h_j| \leq \delta$$
(since initially: $h_i = h_j = 0$).

To maintain this relationship, the two processes will exchange control information. The messages relating to these exchanges will be transported by channels assumed to behave in the following way: the messages can be duplicated, desequenced or lost. As in the previous chapter, however, the channels are assumed to be fair (which means in practice that if the messages are transmitted continuously along a channel, at least one of them will reach its destination).

After giving a solution to the problem, it will be generalized to the case of n processes, which means providing a distributed control algorithm maintaining the relationship:

$$i, j \in 1 .. n : |\ h_i - h_j\ | \leq \delta$$

This solution is then studied in the case of several topologies of the network connecting the processes. This will provide information on the effect of the topology on control, and the solutions available.

2.2. The two-clock case

2.2.1. System structure

The system is made up of two logical sites, with exactly one process on each site (from now on we shall make no distinction between a site and the associated process). The control algorithm in each site can be defined by:

—an interface made up of an access point allowing the process to obtain (successive) clock values— the header is:

function date returns integer

—an implementation of the control ensuring that the relationship between the clocks is maintained.

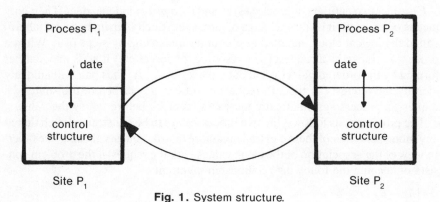

Fig. 1. System structure.

This is a classic breakdown into layers or levels: provision of a service (control algorithm) and use of the service (process).

2.2.2. A solution

2.2.2.1. PRINCIPLE

To have any practical use, the solution must ensure the greatest possible autonomy to the user processes. A simple way of responding to this objective is to dissociate the taking of the clock value from the maintenance of the relationship between the clocks. This involves introducing into each process P_i a local variable max_i which indicates the value to which it can increase its clock h_i at any moment without violating the inequality. The condition associated with the call to *date* is therefore $h_i < max_i$.

The two sites must now exchange the values, to ensure the updating of the variables max_i and max_j such that the system is not blocked. The controller associated with each site can transmit, to this end, the value of the clock it manages. Since this value on the one hand changes while on the other it can be lost during transmission, the controller must transmit this value fairly regularly. The sequence of values transmitted by a controller is therefore a sequence of monotonically non-decreasing integers.

Now consider the reception of these values. The controller receives a sequence of integers formed from a subset of the values sent (since some may be lost), in which certain values may appear several times (either because they have been transmitted several times or because they have been duplicated by the line), and the order in which messages arrive may not be that in which they were transmitted (desequencing is possible).

The sequence of successive limits that h_i can reach without violating the inequality must increase—h_i can only grow. The successive values that the variable max_i can take must form this type of sequence. If the sequence of successive values of the clock h_j received by P_i were monotonically increasing, the following update would respect this constraint while allowing the system to progress (property of liveness):

$$max_i := x + \delta$$

where x is the last value received by P_i. This type of updating would maintain the invariant, giving:
- $h_i \leq max_i$ (by the condition associated with *date*);
- $x \leq h_j$ (by construction of site P_j and non-zero transfer delay)

from which can be deduced:

$$h_i \leq h_j + \delta$$

(The same applies on site P_j.) One way of solving the problem is therefore to consider a sequence of monotonically increasing values received on each site. This involves eliminating the clock values that have been overtaken or duplicated, which is easy since an overtaken value will be lower than the last value received. The variable old_i, representing the last value received, and x, the value being received, are used in the test $old_i < x$ which means that only those successive values received that form an increasing sequence are considered. The problem of desequencing is therefore solved by eliminating desequenced values—that is, by simulating their loss, which is normally resisted by the algorithm.

The site P_i executes the following procedure on reception of value x:

if $x >$ *old$_i$* **then** *old$_i$* $: = x$;
$$max_i : = x + \delta$$
end if

This can be simplified by elimination of the additional variable *old$_i$*, and becomes:

if $x + \delta > max_i$ **then** $max_i : = x + \delta$ **end if**

which, with the addition of the function *maximum*, is expressed as:

$$max_i : = maximum\,(max_i, x + \delta)$$

2.2.2.2. THE CARVALHO–ROUCAIROL ALGORITHM

The above control algorithm allowing the progress of the two clocks to be made interdependent can be expressed by the following simple statements.

Each of the controllers associated with the sites P_i and P_j, with $i = 1$ or 2 and $j + 3 - i$, is equipped with a context formed of the following local variables:

var $h_i : 0 \ldots + \infty$ *increasing initialized to* 0;
　　$max_i : \ldots + \infty$ *increasing initialized to* δ;

and obeys the following behaviour rules:

> **on a call to date returns integer**
>　**possible only if** $h_i < max_i$;
>　**do** $h_i := h_i + 1$;
>　　**return** (h_i);
>　**end do**
> **on transmission**
>　*– – always possible and carried out regularly*
>　**do**
>　　**send** (h_i) **to** P_j
>　**end do**
> **on reception of** (x) **from** P_j
>　**do**
>　　$max_i := maximum\,(max_i, x+\delta)$
>　**end do**

Proof

It is easy to prove this control algorithm. The predicate $\mid h_1 - h_2 \mid \leq \delta$ is true initially; the proof involves showing, by induction, that all operations associated with the events that can arise leave it true. By construction, the sequence of values transmitted by P_i and considered by P_j in the updating of max_j is increasing; let x_i and x_j be the last of these values. So:

$$x_i \leq h_i \leq max_i \qquad \text{(operations } date \text{ and sending of the}$$
$$x_j \leq h_j \leq max_j \qquad \text{clock values)}$$

from which can be deduced:

$$h_i - h_j \leq max_i - x_j$$
$$h_j - h_i \leq max_j - x_i$$

The updating processes for max_i are such that:

$$max_i = x_j + \delta$$

from which it is concluded that the predicate is always true.

2.2.3. A variant

2.2.3.1. THE ASSUMPTION OF NON-DESEQUENCING

Consider the same problem, with modification of one assumption: the messages can be lost and duplicated, but not desequenced. There is therefore implicit information in every message to the effect that all messages transmitted previously have either arrived or been lost. In other words, the sequence of values received by a site is always non-decreasing. The desired property constructed in the previous solution is now provided by the channels. The question that arises is: what is the effect of this property on the algorithm? It will be seen that it allows a reduction in the information exchanged by the messages.

2.2.3.2. RANGE OF VALUES CARRIED BY THE MESSAGES

The processing carried out by P_i during reception of the clock value x is simplified to:

on reception of (x) from P_j
do max_i : $= x + \delta$ end do

we have:

$$x - \delta \leq h_i \leq x + \delta$$

because:

$-h_i \leq x + \delta$ as $h_i \leq max_i$ and max_i cannot decrease
$-h_i \geq x - \delta$ as $h_j \geq x$ and $h_j - h_i \leq \delta$

which can be formulated in the following way: when P_i receives a value x, this differs from h_i by δ at most and can therefore only be one of the $2\delta + 1$ different values in an interval defined by h_i and δ:

$$h_i - \delta \leq x \leq h_i + \delta$$

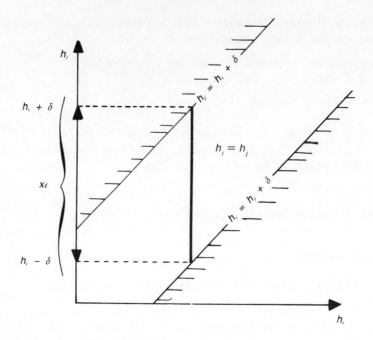

The domain of definition for a message value x is perfectly defined as a function of δ and h_j (since there is no desequencing) when the message is received by P_i, so it is possible to communicate only the **modulo** k values with $k \geq 2\delta + 1$ (size of the range of definition for x).

When P_i receives a value x it must 'demodulate' it to update its variable max_i by calculating the number m such that:

$$h_i - \delta \leq k.m + x \leq h_i + \delta$$

2.2.3.3. THE ALGORITHM

The control algorithm associated with each site is similar to that found previously. In addition, the following declarations are included:

constant $k = 2\delta + 1$;
var $m : 0 .. +\infty$;

The processing carried out during a call to *date* is not modified. For the two other events, the statements become:

on transmission
    do send $(h_i$ **mod** $k)$ **to** P_j **end do**
on reception of (x) **from** P_j
    $-- \times \in 0 .. k-1$
    do

$m := <$ *value such that* $: h_i - \delta \le km + x \le h_i + \delta >;$
$max_i := km + x + \delta$
end do

2.2.3.4. IMPORTANCE

The non-desequencing of messages allows their size to be bounded, since the values transported are included between 0 and $2\delta + 1$. This involves a slightly more costly processing at the two sites ('modulate and demodulate'), but allows the load on the lines to be reduced, which is often important for high clock values: $\log_2 (2\delta + 1)$ bits are sufficient to represent any value transported by the control messages.

2.2.4. Desymmetrizing the problem

This control algorithm is important for another reason. Consider two values δ_1 and δ_2, with desymmetrization of the two controllers associated with the sites P_i and P_j (they will therefore no longer have the same programs), such that they maintain the following invariant relationship:

$$\delta_1 \le h_i - h_j \le \delta_2$$

The new controllers are easily obtained from the old ones:
—the one for site P_i ensures $h_i \le h_j + \delta_2$
—the one for site P_j ensures $h_j \le h_i - \delta_1$
An interesting case arises if $\delta_1 = 0$, $\delta_2 = n$ are chosen:

$$0 \le h_i - h_j \le n$$

It will be seen that this relationship is none other than that of the producer–consumer problem with a maximum gap of n messages : h_i represents the number of productions and h_j the number of consumptions carried out (see Section 1.1.2).

To a certain extent, the clock problem generalizes the producer–consumer problem since it allows a more general solution to be given for problems of mutually dependent control.

2.3. The *n*-clock

2.3.1. Principle of generalization

The problem of mutual control of two clocks is characterized by the perception that each site has of the value of the clock on the other site. It is interesting to generalize this problem because, in addition to the practical importance of the solution obtained, it can provide the elements of generalization methods usable in other problems. Consider the problem in which each of the *n* sites, numbered from 1 to *n*, is equipped with a logical clock. Control must be exercised over the increments of these clocks such that the following predicate is always true:

$$\forall i,j \in 1 \ .. \ n{:}\mid h_i - h_j \mid \ \leq \delta$$

To the assumptions concerning the behaviour of the channels (which are the same as previously: duplications, desequencing and loss possible), another should be added on the structure of the channels. A fully connected network is assumed, with any site being able to communicate directly with any other.

Note that the approach consisting of the establishment of a hierachy between the sites, with a 'master' clock and the others acting as 'slaves', and the simulation of centralized control does not work; each of the clocks must be subject to all the others.

Consider a site P_i. The relationship to be maintained appears in the following form:

$$\forall j \in 1 \ .. \ n : \mid h_i - h_j \mid \ \leq \delta$$

and as all the sites exercise the same control, the controller placed at P_i must simply ensure the maintenance of:

$$\forall j \in 1 \ .. \ n : h_i \leq h_j + \delta$$

To achieve this, P_i must know the 'state' of each of the other sites. Two provisions are necessary for each of the P_i:

—broadcasting: allowing P_i to communicate its successive clock values to each P_j;

—memory: allowing P_i to store the largest clock values communicated by each of the P_j.

With these two provisions, the solution for *n* sites can be derived from the solution for two sites, taking into account the fact that a site interacts with its environment (made up of $n-1$ other sites). Two approaches are possible depending on the use made by a site P_i of the clock value received from P_j: it may use it to construct a copy, albeit delayed, of the value of h_j, or a delayed copy of the maximum value that h_j can attain. Both of these approaches will be examined.

2.3.2. First solution

This is the immediate generalization of the solution with $n=2$. The variable max_i for each of the processes P_i plays the same role. At any moment P_i can broadcast the value of its clock to each P_j; note that this value may be lost on certain channels, duplicated on others and desequenced on yet others, and may arrive correctly on some. This means that two processes P_j and P_k will not necessarily have the same view of the last largest value transmitted by P_i.

A table $last_i [1 \ ... \ n]$ is introduced and used such that $last_i [j]$ stores the last largest clock value received by P_i from P_j. Therefore h_i is constrained by max_i, which must be such that:

—if a single P_j is considered: $max_i \leq last_i[j] + \delta$; therefore:
—if all the P_j are considered: $max_i \leq min \{last_i[j] + \delta\}$
$$i \neq j$$
This is the site at which the progress of the clock is slowest or the clock value is the slowest to be communicated, which slows all the others.

The control algorithm of site P_i is therefore the following:

var $h_i, y : 0 .. + \infty$ *increasing initialized to* 0;
   $max_i : 0 .. + \infty$ *increasing initialized to* δ;
   $last_i : $ *table* $[1 .. n]$ *of* $0 .. + \infty$ *increasing initialized to* 0;

on a call to date *returns integer*
   possible only if $h_i < max_i$;
   do $h_i := h_i + 1$;
      return (h_i)
   end do
on transmission
   – –always possible and done regularly;
   do
      send (h) *to* $P_j : \forall j \neq i$
   end do
on reception of (x) *from* P_j, $\forall j \neq i$
   do
      if $last_i[j] < x$ *then*
         $last_i[j] := x$;
         $y := min \{last_i[k]\}$;
            $k \neq i$
         $max_i := y + \delta$
      end if
   end do

The proof of this algorithm is similar to that in the two-site case. The environment which, at a given moment, constrains P_i is here the site P_j perceived by P_i such that:

$$last_i[j] = min \{last_i[k]\}$$
$$k \neq i$$

This value plays the role of x_j in the previous proof. The proof is therefore made on the basis of the following statement:

$$\forall i, j : min \{last_i[k]\} \leq last_i[j] \leq h_j \leq max_j$$
$$k \neq i$$

and the rest follows as before.

Note: It is certainly possible to improve this solution by not recalculating y when it is certain that it has not been modified by the arrival of x. The procedure therefore becomes (with y initialized to 0):

> *if last$_i$ [j] < x then*
> $z := last_i [j]$;
> $last_i [j] := x$;
> *if y=z then* $y := min \{last_i[k]\}$;
> $k \neq i$
> $max_i := y + \delta$;
> *end if*
> *end if*

A more interesting modification consists of 'homogenizing' the use of the $last_i$ table by eliminating the test $k \neq i$. For this, h_i and $last_i$ [i] must become indistinguishable by making them synonymous or by eliminating h_i after having replaced its occurrences by $last_i$ [i]. Then $last_i$ [i] denotes, according to the semantics of the $last_i$ table, the largest value of h_i known to P_i, which corresponds to the synonymy introduced. The reader is invited to study, in relation to the previous case i) the effects of this modification on the progress of max_i when all the sites have their clocks at k, except that of P_i, which is at $k - \delta$, and ii) the effect of updating max_i during the increase of $last_i$ [i].

2.3.3. Second solution

When P_i receives a clock value for P_j greater than the previous ones, it may no longer consider a delayed copy of the value of h_j but a copy of the value of max_j, the limit attainable by h_j. The perception that P_i has of its environment is based on another viewpoint. For this a table $max_i[1 .. n]$ is introduced at each site, such that $max_i [j]$ defines the limit that h_j can attain as perceived by P_i.

In order to increase its clock h_i, the site P_i:

—(if a single P_j is considered) must not exceed $max_i [j]$ since at worst $max_i [j]=h_j + \delta$, and generally $max_i [j] \leq h_j + \delta$.

—(if all the P_j are considered) must not exceed $min \{max_i [j]\}$.
                                                            $j \neq i$

The updating of $max_i [j]$ is carried out during the reception of a clock value x from P_j by:

> $max_i [j] := maximum (max_i [j], x + \delta)$

(P_j could also send not a value of h_j but its current bound $h_j + \delta$).

The algorithm is as follows:

> *var h$_i$: 0 .. + ∞ increasing initialized to 0;*
>   *max$_i$: table[1 .. n] of 0 .. + ∞ increasing initialized to δ;*
> *on a call to date returns integer*
>   *possible only if* $h_i < min \{max_i [k]\}$;
> $k \neq i$

```
    do
        h_i := h_i + 1;
        return (h_i);
    end do
on transmission
    — always possible and performed regularly
    do
        send (h_i) to P_j, ∀ j ≠ i
    end do
on reception of x from P_j, ∀ j ≠ i
    do
        max_i [j] := maximum (max_i [j], x + δ)
    end do
```

Proof

The proof for this algorithm is easy. The successive values taken by the variables $max_i [j]$ form increasing sequences. Also:

$$\forall \ i, j : h_i \leq \min_{j \neq i} \{max_i [j]\}$$

hence:

$$\forall \ i, j : h_i \leq max_i [j]$$

but $max_i [j] \leq h_j + \delta$ (the value that allows $max_i [j]$ to be calculated is less than or equal to h_j). Thus:

$$\forall \ i, j : h_i \leq h_j + \delta$$

which proves the invariant.

Note: It is interesting to compare the two algorithms given here. In both cases, each site perceives only a previous state of each of the other sites. It is the viewpoint adopted that differentiates the two solutions: the effect of the environmental view adopted by P_i is based on the condition associated with the *date* procedure.

As in the version with two clocks, the range of values transported by messages can be limited in these algorithms if the messages are not desequenced.

2.4. Arbitrary network

2.4.1. Algorithms in an arbitrary network

In the two previous algorithms a site is constrained by its environment, constituted by the neighbour sites. As the hypothesis states that the network is fully connected is complete, the sites adjacent to a given site are, in fact, all the other sites of the network. The following question arises: what control is exercised by either of the previous algorithms if each site is constrained only by its immediate neighbours and if the network is connected but arbitrarily? In other words, if the symmetrical neighbourhood relationship is defined by the following predicate *neighbouring*:

neighbouring $(P_i, P_j) = $ *true iff P_i and P_j are directly adjacent*

Any one of the previous algorithms, without requiring modification, ensures that:

$i, j : neighbouring$ $(P_i, P_j) = > |h_i - h_j| \leq \delta$

What relationship for any two clocks is ensured by these control algorithms placed at the network sites?

Let d be the diameter of the non-oriented graph associated with the network (note that this refers to the longest of the shortest acyclic paths between any two sites). So:

$$\forall \, i, j : |h_i - h_j| \leq d.\delta$$

In the case of a ring, for example, two neighbouring sites have clocks between which the maximum difference is δ, but the difference is $[n/2]\delta$ for any two sites. If the sites are placed in line, the maximum difference (between clocks placed at the end of the line) is $(n-1)\delta$.

In the case where an application requires particular drift values δ_{ij} between the unordered pairs of adjacent sites P_i and P_j, the algorithms can easily be adapted to take this characteristic into account. The maximal difference between any two clocks is then obtained by weighting each edge (P_i, P_j) of the graph by the corresponding value δ_{ij} and calculating the path of minimal weight: the controllers maintain the following inequality:

$$\forall \, i, j |h_i - h_j| \leq \min_{u \in S} \left(\sum_{(P_k, P_l)_{\epsilon u}} \delta_{kl} \right)$$

where S is the set of acyclic paths between P_i and P_j.

2.4.2. Channel breakdowns

2.4.2.1. EFFECTS OF BREAKDOWNS

The assumptions concerning the behaviour of the channels take them to be fair. This means in practice that if a controller constantly sends messages down a given channel, at least one of them will reach its destination. A channel that is cut or that breaks down completely is therefore no longer fair. The effect this has on the algorithm is considered below.

Consider the two systems R1 and R2 shown here, each made up of four sites. Unlike R2, the network for R1 is a complete fully connected one (Figure 1).

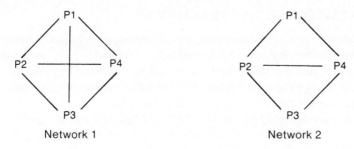

Network 1   Network 2

Fig. 1. Two systems with different behaviour.

In each of these systems, each of the sites P_i is equipped with one of the previous control algorithms, which makes the incrementing of its clock subject to the neighbouring clocks. The relationships maintained are therefore (with the diameters of R1 and R2 being 1 and 2 respectively);

—system R1 : $\forall\, i,\, j |h_i - h_j| \leq \delta$
—system R2: $\forall\, i,\, j |h_i - h_j| \leq 2\,\delta$
—system R1 or R2 : $neighbouring\ (P_i,\, P_j) => |h_i - h_j| \leq \delta$

Now suppose that in system R1, the channel (P_1, P_3) breaks down completely: it therefore does not transmit any more messages and loses its property of fairness. After this breakdown the system R1, now called RB1, has the same topology as system R2, RB1 and R2 have the same communications structure, but do they behave in the same way?

The control carried out on the clocks by RB1 is the same as R1 (except that one line has disappeared); RB1 therefore obeys the same constraints as R1, so RB1 and R2 behave differently. But what is the difference between the behaviour of RB1 and R1, which have the same control algorithms at the same sites but different communication structures? In RB1 the site P_3 (or P_1), once h_3 has reached max_3 (if the formula for the first algorithm given is adopted), must observe an increase of max_3 to advance h_3.

Thus:

$$max_3 = min\ \{last_3\ [j]\} + \delta$$
$$j=1,\, 2,\, 4$$

But $last_3\,[1]$ remains fixed because P_3 does not receive the values sent to it by P_1, since link (P_3, P_1) no longer functions. So max_3 will remain equal to $last_3\,[1] + \delta$ and the clock at site P_3 will not advance. Because of the symmetry of the control algorithms, the same will be true of P_1, P_2 and P_4.

The main result is the following: the complete breakdown of a line (loss of fairness) leads to the system being blocked. It is easy to show that this result is independent of the neighbourhood relationship, integrated into the control algorithms, which defined the initial topology (which can be of any type, as long as it is connected) allowing the sites to communicate.

The next section considers how to resist breakdowns.

2.4.2.2. RESISTANCE TO BREAKDOWNS

The introduction of breakdowns into the communication lines is important for two reasons: it allows the control achieved to be better understood, particularly as regards the perception of the system at each site; and it introduces the question of resistance to breakdowns and how to ensure that the system is not blocked as long as the graph connecting the sites remains connected.

Consider, in system RB1 above, the site P_3 blocked by the absence of messages from P_1, which would increase the value of $last_3$ [1] and therefore max_3. To avoid this deadlock, one solution consists of transporting the desired information to P_3 by other paths than the inoperative channel (P_1, P_3). There are two possible routes: via P_2 or P_4. As messages can be lost other than by channel breakdowns, and the symmetrical character of the algorithm is to be retained, both possibilities must be considered.

One solution consists of providing each of the sites P_i with the following behaviour: when it receives a value it broadcasts it to all adjacent sites. In order that the latter can associate the value received with one of the clocks, each message is made up of two fields: a clock value x and the site k to which it relates. This is simply the systematic dispatch of the values received to the neighbouring sites. On receiving such a message (which comes from or via a neighbouring site) a site may take it into account systematically or only if the clock value transported relates to one of the neighbouring sites. The relationships maintained are therefore respectively that of the complete graph and of the graph of diameter d defined by the relationship *neighbouring*, describing initial adjacence. In the latter case, the behaviour associated with reception of a message is the following:

on reception of (x, k) *from* P_j
  do
    if $k \in neighbours_i$ *and* $last_i[k] < x$
      then $last_i[k] := x;$
            $y := min \{last_i[l]\};$
                $\quad l \neq i$
            $max_i := maximum (max_i, y + \delta)$
    end if;
    send (x, k) *to* $P_l,$ $\forall l \in (neighbours_i - \{j\})$
  end do

The retransmission of the message received must be carried out outside the *if* . . . *then* . . . *fi* loop; if it is done inside, it will only occur once at the most, which will not permit the algorithm to resist message losses. It is not essential to send the message (x, k) to P_k if P_k is directly adjacent to P_i in the case where the message transmitter is not the one adjacent to P_k. The solution can also be improved by propagating not x but the most recently updated value, $last_i[k]$. The last statement performing the transmission of the message therefore becomes:

case $k \in neighbours_i \rightarrow$ ***send*** $(last_i[k], k)$ *to* $P_l: \forall l \in (neighbours_i - \{j, k\})$
      $k \in neighbours_i \rightarrow$ ***send*** (x, k) *to* $P_l: \forall l \in (neighbours_i - \{j\})$
end case

This suggests the dissociation of the reception of a message from its transmission to neighbours. Each time a site P_i transmits a value of its clock, it can also send the most recently updated information it has on the clocks in other sites—in other words, its *last*$_i$ table (in which *last*$_i$[i] is synonymous with h_i, in the interests of homogeneity of expression). This also allows a clear differentiation to be made between the two functions of P_i (transmission and reception) in the overall control exercised. In the case where the relationship maintained relates to the complete grid (d=1), the following procedure is obtained:

> *on transmission*
>   *– –always possible and performed regularly;*
>   *do*
>   *send last$_i$ to P_j,* $\forall j \neq i$
>   *end do*
> *on reception of (last) from P_j,* $\forall j \neq i$
>   *do*
>   *last$_i$ [k] : = maximum (last$_i$ [k], last [k]),* $\forall k$
>   *y : = min {last$_i$ [k]};*
>   *max$_i$: = maximum (max$_i$, y + δ)*
>   *end do*

The resistance of the channels to breakdowns is obtained by the 'systematic' retransmission of the values received to neighbouring sites. The control obtained ensures the functioning of the system as long as the graph connecting the sites remains intact. Any site remains subject, whatever the structure of the system after possible breakdowns, to the clocks of the sites initially defined as being neighbours. It therefore appears that, whatever the initial connections between the sites, the control thus implemented ensures the maintenance of the relationships among logical neighbours.

The resistance to breakdowns obtained is interesting in more than one sense. Firstly, it ensures the (qualitative) non-locking of the system as long as the graph remains connected. Quantitatively, it can improve site performance by reducing periods of temporary locking by the systematic exchanges introduced. Finally, the relationship maintained by the control algorithm is independent of the connections (initial or after breakdown) between sites. It depends only on the direct neighbourhood relationship defined *a priori*. In addition, in the case where the table *last$_i$* is transmitted in the messages, and these are not desequenced, the complexity of communication is $O(n \log_2 (2\delta+1))$ bits.

2.5. References

The problem of logical clocks, as presented by Carvalho and Roucairol (1985) in the case of two clocks, contains suggestions for implementation, particularly that involving bounded messages in the case of non-desequencing. This was generalized to n clocks by Raynal (1986), who also considers networks with arbitrary topology.

Techniques for systematic broadcasting of messages received in order to resist breakdowns have been proposed by various authors, notable among whom are Tajibnapis (1977) and Segall (1983), who illustrate this technique in protocols for maintaining the coherence of network routeing tables.

Two fundamental elements: elementary algorithms and global state

This part of the book tackles two basic elements which, in the general case, are necessary to the design and realization of distributed applications and systems. These really concern a certain number of simple algorithms allowing traversal through a network, learning of the structure, calculation of the routes, etc. These algorithms concern the network part of the applications and make it possible to relax the initial conditions such as the need for knowledge of the number of sites, for example. The second aspect concerns the system part of the applications. For a given site, this means the detection of a global state allowing the site to take a decision or verify a property.

Depending on the system or application implemented, these two aspects must be resolved with wholly or in part, and it is important to know how to solve them.

Chapter 3

Basic algorithms and protocols

3.1. Introduction

The transfer protocol studied in Chapter 1 involved a system made up of two sites; the underlying communication network allowed each of them to communicate directly with the other, and the graph modelling this network was therefore complete. The control exercised by the algorithm introduced in Chapter 2 was based initially on a complete network; then, in order to resist breakdowns of communication channels, the algorithm was modified; finally, the communication structure necessary to its functioning was shown to be an arbitrary network whose topology can change as a function of channel breakdowns, while its graph remained connected.

The graph modelling the communication network of a distributed system or application can, in the general case, be of any type as long as it is connected (in other words, there is one and only one system). In the case where the network conforms to this, it may be necessary to use a certain number of basic algorithms to relieve the load on the algorithms concerned with the distributed application of certain functions. This means providing one or more layers which would carry out the network function, allowing each site to avoid handling certain problems of information transfer and the associated control.

The aim of this chapter is to present such algorithms (or protocols) in the case of information routing problems. The functions carried out (also called 'services rendered') by these protocols are essential because they can, if necessary, make the implementations independent of the network graph in which they appear. The protocols introduced therefore allow:

—traversal through the network according to various strategies;

—the learning of global information not initially known by the sites, such as the structure of the network itself, for example;

—the determination of the optimal route between sites;

—the updating of the communication routes if the channels can break down and be repaired.

Independently of the functions carried out, some of the basic protocols are of great importance: they are based on techniques of distributed computation which are at the basis of many other distributed algorithms. They constitute protocols with a dual importance: because of the function carried out and because of the algorithmic structure shown.

3.2. Assumptions

The algorithms introduced are defined in a certain context. Apart from specific assumptions mentioned separately, this general context is defined by the following characteristics.

3.2.1. Assumptions concerning the network

The network is modelled by a graph G, which may be of any form as long as it is connected, in which the vertices X correspond to the set of sites, and in which the set of edges is made up of the channels. These are bidirectional (G is non-oriented).

The channels are reliable; they do not alter the messages, lose them, duplicate them or desequence them. Any message transmitted along a channel by a site will reach the site to which the channel connects it after an arbitrary, but finite, period of time.

3.2.2. Assumptions concerning initial knowledge

The sites all have distinct names and know that their names are distinct. The names are taken from the integers 1, 2, . . . *maxn* (it is therefore possible to order the sites strictly according to their identities). Certain protocols require that the sites initially know *maxn*, the maximal number of sites that can be present in the system, although there can be fewer. No site initially knows the structure of the network; they know only the channels connecting them to their neighbours, whose identity they do not necessarily know.

3.2.3. Importance of these assumptions

The consideration of an arbitrary graph and the initial lack of knowledge of this graph by any of the sites are two interesting hypotheses. The modification of the system structure by the addition or suppression of sites does not require the entire system to be redefined: only the new sites and those where the adjacency relations have been modified must be redefined (and therefore re-compiled). This is of particular interest for the generation of distributed systems or for re-establishment after breakdowns. From this point of view, one algorithm is better than another if the knowledge required at each site is as local as possible. The two hypotheses given here fulfil this criterion.

3.3. Traversal of a network

Many applications require traversal of messages through the network supporting the distributed system. The problems of traversal of a graph have been studied in full in a sequential context, but here the same aim is to be achieved in a distributed context.

3.3.1. Depth-first traversal

3.3.1.1. THE PROBLEM

Consider a site P_r. The aim is to equip the set of sites in the system with a control algorithm distributed over these sites, which will allow P_r, using messages, to effect a 'depth-first' traversal through the graph associated with the system.

An algorithm of this type can be used as the component of a more general protocol; consider, for example, the broadcasting of information to all the sites, the assignment of different names to the sites, the calculation of a tree-structure covering the network associated with a pre-determined root, etc.

3.3.1.2. PRINCIPLE OF THE SOLUTION

The principle on which the algorithm is based is that of site marking. When a site receives a control message for the first time, it marks itself and then sends this message on to its neighbours. When an already marked site receives a message, it immediately replies to the transmitter that it has already received a message. A site also responds to the site that caused it to be marked when it learns that all its neighbours are marked. The site P_r which started the traversal knows that the calculation has come to an end when it is informed that all its neighbours are marked (in fact the neighbours to P_r's neighbours are then also marked, and so on).

At the end of the traversal the initiator P_r is the root of a tree-structure, including all the sites and such that a given site has as its predecessor the site which led to its being marked.

3.3.1.3. THE ALGORITHM

The messages are therefore of the types *scan* and *backtrack*. They carry the identity of the transmitter site.

Each site P_i is equipped with the following definitions:

var marked$_i$: *boolean initialized to false*:
  neighbours$_i$, aux$_i$: *set of names initialized to* {*names of neighbours to P_i*};
  parent$_i$: *non-negative integer*;

The variables *neighbours$_i$* show the adjacency relationship that links the sites. Although the channels are bidirectional (the messages scan and backtrack in opposite directions), the variables *neighbours$_i$* can indicate an oriented or non-oriented graph. In the former case the adjacency relationship is not necessarily symmetrical and only the *scan* messages move in the direction of the arcs. (Moreover, only the sites accessible from P_r will be visited.)

To simplify the writing of the algorithm, the position of a site is defined by the identities of its neighbours. This hypothesis is not necessary to the correct functioning of the algorithm; the distinction between the channels connecting it to its neighbours is sufficient. Instructions must therefore be added to the algorithm to allow it to learn the names of interlocutory sites during communication, and to draw up a table of correspondences that it can later use.

The additional variable aux_i stores the neighbours from which P_i is sequentially expecting *backtrack* messages. The variable *parent$_i$* stores the predecessor of P_i, that is the first site that led to its being marked.

The algorithm is given here for the site P_i. All the sites have the same algorithm, with the single exception of P_r, the traversal initiator.

> **on decision to initiate a traversal**
>   **possible only by P_i such that $i=r$,**
>   **do** *marked$_r$* : = *true*;
>   *aux$_i$* : = *neighbours$_r$*;
>   *x* : = *an element (aux$_i$)*;
>   **send** *scan* (*r*) **to** P_x
>   **and do**
> **on reception of** *scan* (*j*)
>   **do**
>   **case** *marked$_i$* → **send** *backtrack* (*i*) **to** P_j
>   ¬*marked$_i$* → *marked$_i$* : = *true*;
>   *parent$_i$* : = *j*;
>   *aux$_i$* : = *neighbours$_i$* −{*j*};
>   **case** *aux$_i$* =∅→**send** *backtrack* (*i*) **to** P_j
>   *aux$_i$* =∅→*x* : = *an element (aux$_i$)*
>   **send** *scan* (*i*) **to** P_x
>   **end case**
>   **end case**
>   **end do**
> **on reception of** *backtrack* (*j*)
>   **do**
>   *aux$_i$* : = *aux$_i$* − {*j*};
>   **case** *aux$_i$* = ∅ **and** *i=r* → *traversal ended*
>   *aux$_i$* = ∅ **and** *i≠r* → *x* : = *parent$_i$*; **send** *backtrack* (*i*) **to** P_x
>   *aux$_i$* ≠ ∅   → *x* : = *an element (aux$_i$)*;
>   **send** *scan* (*i*) **to** P_x
>   **end case**
>   **end do**

3.3.1.4. EVALUATION

It is easy to show that:
—the sites accessible from P_r are marked once and once only;
—the algorithm ends.

The termination is known to P_r: if necessary, this site can propagate a message (using the existing tree-structure) through the system to inform the others that the process has ended.

The complexity of a distributed algorithm can be evaluated using various measures, among which the size and number of the necessary messages are of the greatest importance. A message belongs to one of two possible types and transports a value lying between 1 and n. Hence $1 + \log_2 n$ bits are necessary to represent a message. As for the number of messages, it will be noted that for every *scan* message transmitted along a channel, there is a corresponding message *backtrack*, transmitted in the opposite direction. Each site sends a *scan* message to each of its neighbours, except for the one that caused its marking to occur. Each edge transports at the most one *scan* message in each direction; if the graph has e edges, the algorithm will activate $O(e)$ messages. If the two results are combined, the communication cost of the algorithm is: $O(e. \log_2 n)$. (Note that, at the worst, e is equal to $n(n-1)$ in an oriented graph and to $n(n-1)/2$ in a non-oriented graph.)

3.3.1.5. USES

As explained, this algorithm can constitute the skeleton on which other algorithms can be based, because of the sequential traversal it achieves for the network. Note particularly, as regards these other algorithms:

—the initialization of systems: using it to give distinct identities to the various sites (starting from a version based on the channel names);

—the calculation of maximum flow in a network (distributed version of the Ford–Fulkerson algorithm);

—the study of reliability: using it to find the articulation points and the bi-connected components of the network, or more generally to calculate the connectivity (note that a network is k-connected as long as the removal of $k-1$ arbitrary sites does not divide it into two or more networks).

3.3.2. Parallel traversal

3.3.2.1. THE PROBLEM AND PRINCIPLE OF ITS SOLUTION

This is the same problem as before, but it is not the case that traversal of the network must take place by the sequential visiting of all the sites; several sites can be visited simultaneously by control messages. This is parallel traversal.

A simple solution consists of applying the principle of flooding: when P_i first receives a *scan* message, it marks itself and sends the message to each of its neighbours with the exception of the original transmitter of the message. After this P_i will not react to subsequent reception of the *scan* message. There is, therefore, only one type of message and P_i's algorithm is basically:

on reception of scan *(j)*
   do
      if ¬ *marked$_i$ then*
              marked$_i$: = *true*;
              parent$_i$: = *j*;
              aux$_i$: = *neighbours$_i$* − {*j*};

> **if** $aux_i \neq 0$ **then**
> 　　$\forall\, x \in aux_i$: **send** *scan* (i) **to** P_x
> 　　**end if**
> 　**end if**
> **end do**

The traversal is started by P_r by sending the message *scan*(r) to all its neighbour sites.

This algorithm ensures the propagation of *scan* messages through the whole network; during this propagation any line transports one or two *scan* messages (in opposite directions). The complexity of the number of messages exchanged is therefore $O(e)$.

Although correct, this algorithm is not satisfactory. Site P_r, which initiates the parallel traversal, does not know when the traversal is ended. The only knowledge it has is that the traversal through the graph will finish. In the previous algorithm (traversal in depth) the initiator site P_r knew when the traversal was ended: this knowledge came to it by the predicate $aux_r = 0$ changing to true, indicating that all its neighbours (and therefore their neighbours) had been visited.

This information on the ending of the algorithm may be essential for algorithms based on such traversal. The broadcasting of information by site P_r to the other sites, based on this search (the message *scan*, transports this broadcast information as well as the identity of the transmitter) can, for example, require an acknowledgement which, once received by P_r, indicates to it that each site has received the information. This means offering P_r the knowledge of the ending of the search when it occurs.

To achieve this, for every *scan*-type message m there will be a corresponding *acknowledge*-type message which moves in the opposite direction, to indicate that the message m has arrived. When a site has received all the acknowledgements relating to the messages transmitted in parallel, it knows that they have arrived. It is therefore important to place the sending of the acknowledgement messages correctly so that, when a site has received all the messages expected, it knows that the traversal has ended in the subnetwork in which it has played the role of initiator. This check is easy to achieve: each site must refrain from acknowledging the first *scan* message it receives until it has received all the acknowledgements it was expecting. It can acknowledge the other *scan* messages at any time—for example, as soon as it receives one.

P_r will therefore learn, with the last acknowledgement message it receives, that the parallel traversal through the network has ended. (The check used to obtain knowledge of this end is none other than a suitable use of 'parentheses', with *scan* and *acknowledge* respectively playing the parts of open and close parentheses.)

3.3.2.2. THE ALGORITHM

In the traversal-in-depth algorithm, only the parent of each site in the constructed tree was given. Here, partly because of its potential usefulness and partly because of its importance from the point of view of distributed computation, the predecessor and successors relating to the parallel traversal will be calculated for each site. The context of P_i is similar to what it was before:

> *var marked$_i$: **boolean initialized to false**;*
> *neighbours$_i$: **set of names initialized to** {names of neighbours*
> *of P_i};*
> *succ$_i$: **set of names**;*
> *parent$_i$: 1 .. n;*
> *nback : **non-negative integer initialized to** 0;*

The variable *succ$_i$* will define the successors to the site P_i; *nback$_i$* gives the number of *acknowledge* messages expected before the sending of the last *acknowledge*. When P_i receives the first *scan* message, it learns who its predecessor is. It then propagates *scan* to all its neighbours (except the predecessor from which it received it), which it considers as its successors. But a site may receive *scan* messages from several of its neighbours: it will then be the successor to each of them, but if the tree-structure is to be consistent it can only be the successor to its one predecessor. To solve this problem the messages *acknowledge* will be validated by *alreadyseen* or *ended*; the message *acknowledge (alreadyseen, i)* will be sent by P_i to acknowledge a *scan* message when it has already been visited: the receiver of such a message will remove P_i from its successors; *acknowledge (ended, i)* will be used by P_i to acknowledge the first *scan* message received and it will therefore validate P_i as successor.

> *on **decision to** initiate traversal*
> *possible only by P_i such that i=r;*
> *do mark$_i$:= true;*
> *succ$_i$:= neighbours$_i$;*
> *nback$_i$:= cardinal (succ$_i$);*
> *∀ x ∈ neighbours$_i$: **send** scan(r) to P_x*
> ***end do***
> *on **reception of** scan(j)*
> ***do***
> *case marked$_i$ → **send** acknowledge(alreadyseen, i) to P_j*
> *¬marked$_i$ → marked$_i$:= true;*
> *parent$_i$:= j;*
> *succ$_i$:= neighbour$_i$ −{j};*
> *case succ$_i$ = 0 → **send** acknowledge(ended, i) to P_j*
> *succ$_i$ ≠ 0 → nback$_i$:= cardinal(succ$_i$);*
> *∀ x ∈ succ$_i$: **send** scan(i) to P_x*
> ***end case***
> ***end case***
> ***end do***
> *on **reception of** acknowledge (x,j)*
> ***do***
> *if x = alreadyseen then succ$_i$:= succ$_i$ − {j} **end if**;*

$$nback_i := nback_i - 1;$$
if $nback_i = 0$ **then**
   case $i = r \rightarrow$ *traversal ended*
      $i \neq r \rightarrow x := parent_i;$
            send *acknowledge(ended, i)* **to** P_x
   end case
   end if
end do

3.3.2.3. COMMENTS

The formation of a tree-structure is not necessary to a parallel traversal of the network; it was shown here for the reasons given above. A construction of this type can also be made in the case of traversal in depth: the messages *backtrack* must then be validated to allow their receiver P_i to retain the transmitted P_j in, or remove it from, the set of successors $succ_i$. As an exercise, the reader is invited to restructure the traversal in depth, taking into account the $succ_i$ variables.

Different executions of parallel searches issuing from a single initiator P_r can give different tree-structures with root P_r. In fact, no assumption is made concerning the speed of the messages; these speeds can differ from one traversal to another, since the sites may not receive the same *scan* messages in the same order at each execution: different tree-structures therefore result. For a given execution, let t_{ij} be the time taken for the transfer of the message *scan* from P_i to P_j, and let the graph edges associated with the network be weighted by these values; the tree-structure formed by one traversal is the tree for the paths of minimum value between P_r and any other site. The traversal obtained is optimal in that direction.

3.3.2.4. EVALUATION

The complexity in number of messages of this algorithm is the same as before: each channel transports at the most one *scan* message in each direction. Associated with each *scan* message is an *acknowledge* message. The complexity is therefore $O(e)$, where e is the number of channels.

Without modifying the complexity, it is possible to reduce this number by considering only the *scan* messages that travel once and once only in each direction on each line: in one direction they have their normal semantics; in the other, that of an acknowledgement. The corresponding algorithm may be derived as an exercise.

The complexity in terms of time of the algorithm is proportional to the diameter d of the network; if Δ is the maximal transfer time for a message on a line, it is $O(d\Delta)$.

3.3.2.5. USES

Simplified versions of the algorithm presented here have been used as skeletons for various distributed algorithms. This will be discussed in the following chapters in connection with particular problems, for example:

—the broadcasting of information;

—problems of site election and determination of the associated routing;

—calculation of properties in distributed systems, such as the detection of deadlock due to the communications, and termination.

—the definition of optimal routes in a network and the determination of bi-connected components.

3.3.3. Generalizations

The two distributed network traversals considered above are carried out from a site P_j; the tree-structures constructed therefore have this site as their root. The algorithms introduced can easily be generalized such that any site can initiate traversal: each site P_i must then be equipped with additional variables which will store its position in the tree-structure constructed by the traversal originated by a site P_j.

Two solutions are possible: using arrays or lists. The entry k of the arrays for P_i will be relative to a given initiator P_k. These storage structures must be independent of the number n of effective sites in accordance with the hypothesis of absence of global knowledge. If the application is based on lists there is no problem; if it is based on arrays, it is sufficient to know a largest value for the number of sites, *maxn*. Each site P_i is therefore provided with the variables:

> **var** *marked$_i$* : *array* [1 ... *maxn*] *of boolean initialized to false*;
>   *parent$_i$* : *array* [1 ... *maxn*] *of names*;
>   *succ$_i$* : *array* [1 ... *maxn*] *of names*;
>   *nback$_i$* : *array* [1 ... *maxn*] *of* 0 ... *maxn*;

The messages must now carry an additional piece of information: the name of the site that initiated the traversal they are carrying out. Thus a message transmitted by P_j relating to a traversal initiated by P_k is presented in one of the following forms: *scan* (k, j), *backtrack* (k, j) or *acknowledge*(k, x, j) according to the type of traversal.

On receiving such messages, the site P_i reacts according to the type of message and to the value of k which identifies the corresponding traversal. The algorithm of the corresponding parallel traversal thus becomes exactly the same for each of the sites P_i.

> **on decision** *to initiate a traversal*
>   **possible only if** ¬ *marked$_i$* [*i*];
>   **do**
>   *marked$_i$* [*i*] := *true*;
>   *succ$_i$* [*i*] := *neighbours$_i$*;
>   *nback$_i$* [*i*] := *cardinal* (*succ$_i$* [*i*]);
>   ∀ *x* ∈ *neighbours$_i$* : **send** *scan* (*i*, *i*) *to* P_x

```
        end do
on reception of scan (k, j)
    do
        case marked_i [k] →send acknowledge(k, alreadyseen, i) to P_j
            ⌐marked_i [k] → marked_i [k] := true;
                          parent_i [k] := j;
                          succ_i [k] := neighbour_i −{j};
                          case succ_i [k]=0 →send acknowledge(k, ended, i)
                                                                    to P_j
                               succ_i [k]≠0 →nback_i [k] :=cardinal(succ_i [k]);
                                             ∀x∈ succ_i [k]: send scan (k, i) to P_x
                          end case
            end case
    end do
on reception of acknowledge (k, x, j)
    do
        if x = alreadyseen then succ_i [k] := succ_i [k]−{j} end if;
        nback_i [k] :=nback_i [k] − 1;
        if nback_i [k] = 0 then
            case k=i →traversal started by P_i ended
                 k≠i → x := parent_i [k];
                              send acknowledge(k, ended, i) to P_x
            end case
        end if
    end do
```

This algorithm can easily be deduced from the basic model. The same is true for many protocols that require a traversal through the network (for example, the detection of deadlock or termination).

3.4. Learning the network

3.4.1. Learning problems

The hypothesis has been made that the sites have no *a priori* knowledge either of the number n of effective sites making up the system or of its structure (which may be of any type). In the previous generalization using an array, each site knows *maxn*, a maximum value of n, but the same generalization based on a list does not require this information (in exactly the same way as the procedure *factorial(n)* does not require the knowledge of a maximum value for n).

In certain applications a site may need to know the number of participating sites, their effective identity and even the global structure of the network. One solution consists of incorporating this global information into the definition of the sites. This is practicable and constitutes a good solution as long as the information does not change: this is not the case when the global information can change, for then each modification to any item of global information requires the redefinition of each of the sites and therefore a new global definition of the distributed system or application. One possible solution would be to place no global information in the initial context of the sites, and to provide them with a distributed algorithm (executed, for example, at the initialization of the system) which would allow them to learn the desired global information. These are known as learning algorithms. The importance of such algorithms is clear: a new definition of the system (addition, modification, suppression of sites, modifications to the network, etc) would require the (re)-definition only of new sites, or of those in which the context (internal or communications) is modified: the other sites would not need to be redefined even though n or the structure of the network had been changed.

Many learning algorithms are possible. Two such will be considered, one for the learning of identities and of n, the other for the learning of the network structure.

3.4.2. Learning the identities

3.4.2.1. IDENTITIES OF NEIGHBOURS

Each site P_i is assumed to have a unique identifier i in the system. Each site therefore has *a priori* some global knowledge: its own identifier is different from that of the other sites. It has been shown that it is possible to carry out a depth-first search from a given site P_r without using the identifiers of the sites, and that this sequential search can be used to advantage for the attribution of the distinct site identities. This allows the assumption to be justified.

A preliminary piece of information that may be necessary to a site is the identifier of its neighbours. This is easily achieved with a protocol for exchange of identity between any pair of sites connected by a bidirectional channel. One or more sites can decide to carry out such an exchange; and this can be generalized to the whole network.

Each site P_i is provided with the following context:

var participant$_i$: *boolean initialized to false*;
  channels$_i$: *set of lines initialized to{links from P_i to other sites}*;
  neighbours$_i$: *set of names initialized to* 0;

The variable *participant$_i$* indicates that if P_i is participating in the exchange of identities, *channels$_i$* constitutes the only information P_i has about neighbourhood relation. The function *associate* allows a one-to-one correspondence or 'bijection' to be set up between an element of *channel$_i$* and an identity of *neighbours$_i$*.

Each *identity*-type message transports the identity of the site that transmitted it. (Although all the messages are of the same type, a name has been given to this type to make the explanations clearer.)

With these elements the algorithm executed by each site P_i is:

> *on decision to initiate an exchange*
>   **do**
>   *participant$_i$* := **true**;
>   ∀ *c* ∈ *channels$_i$* : **send** *identity*(*i*) on *c*;
>   **end do**
> *on reception of identity*(*j*) **on** *cl*
>   **do**
>   *associate* (*cl*, *j*);
>   **if** ⌐ *participant$_i$* **then**
>   *participant$_i$* := **true**;
>   ∀ *c* ∈ *channels$_i$* : **send** *identity*(*i*) **on** *c*
>   **end if**
>   **end do**

Note that the general assumptions concerning the network (Section 3.2.1.) specify that the messages are not lost. If such a stipulation were not fulfilled, it would be necessary to introduce systematic retransmissions until the acknowledgement of reception of a transmitted *identity* message had arrived. The reader is recommended, by way of exercise, to describe such an algorithm (cf. Chapter 1).

Under the general assumptions, two *identity* messages are transported in opposite directions on each link: the complexity of the number of messages is therefore 2*e* where *e* is the number of lines. In this algorithm any site also knows when the exchanges that concern it have ended: it will have received a message on each of its communication links. If it wishes to know when all the exchanges have ended, it can use a depth-first scan modified in the following way: a site that receives a *scan* message waits to learn the identity of all its neighbours before allowing it to progress.

3.4.2.2. IDENTITIES OF ALL THE SITES

When a site knows the identity of all the other sites, it also knows their number. For this reason, only the learning of the other site identites by P_i will be considered here.

A preliminary solution is based on the algorithm of the parallel network traversal (if the identities of its neighbours are not known, it is reformulated by replacing them with the names of the communication links, or the previous learning algorithm is executed). Each site P_i can use this type of algorithm (with P_i as the initiator of the traversal) to broadcast its identity to all the others. On reception of the first message relating to such a diffusion, a site broadcasts its own identity, if this has not already been done. According to our assumptions, the messages are transmitted in a finite time, so each site will know the identities of all the other sites after a finite period of time. One question, however, remains unanswered: at a given moment a site knows that the algorithm *will* stop, but does not know whether it has *yet* stopped—that is, whether it has yet received all the identities.

Answering this question involves adding the following to the learning algorithm: a given site P_i learns that (the identity learning algorithm is ended and that therefore) it will not learn anything more.

The solutions to this problem are based on the following observation: for the site P_i, knowing the identities of the other sites is a question of knowing the identities of its neighbours, then the identities of their neighbours, etc. It will therefore know all the identities when, on learning the identities of the neighbours to the sites it has just learnt, it realizes that it knows them all already. P_i then knows that it has nothing more to learn. This is a closure problem well known in graph theory.

Any message must transport the identity of a site and those of its neighbours to allow any site P_i to test locally for the above global property, which, if found true, will indicate to it that it knows the identities of all the sites. In addition to this knowledge, P_i then knows the structure of the network since it knows all the sites and all their neighbours (if it has retained a copy of the various messages received). For a given site P_i, in view of the property of connection in the network, learning the identities of the other sites and knowing that it knows them all come down to learning the topology of the network.

Note: In the specific case where each site knows the number n of sites in the network, a site knows that it will learn nothing more once it has learnt n distinct identities. This solution is based on global information known *a priori* to all the sites: if this information is not known to P_i, the previous principle must be used.

3.4.3. Learning the network

To simplify the algorithm, consider the case where each site knows its direct neighbours by their identities (either initially or after learning). Learning the network, for a given site, means learning the associated graph G—that is, the set of sites (vertices of G) and the set of links (edges of G).

One or more sites can, *a priori*, initiate the learning algorithm simultaneously. For this a site which has not started it will do so, in the way appropriate itself, as soon as it receives a message, and before processing this message. Only one type of message is necessary, which carries the identity of a site j and the set v of identities of its neighbours, *identity (j, v)*. When a site P_i receives a message of this type from its neighbour P_k, it verifies that it has not already received this message from another neighbour; if this is the case, it updates the set of sites of which it knows the neighbours (by their identities), adding j and the set of connections with j and the elements of v that it knows; it then retransmits the message received to its neighbours other than P_k.

The previous elements make it necessary to equip each site P_i with the following procedure:

*var participant$_i$: **boolean initialized to false**;*
　*neighbours$_i$: **set of names initialized to** {identities of neighbours to P_i}:*
　*knownsites$_i$: **set of names initialized to** {i};*
　*knownchannels$_i$: **set of pairs of names init to** {(i, x):x∈neighbours};*

P_i knows that it has learned all the network when all the channels it knows have both ends in the set of sites that it has learned. It has then received a message *identities(j, v)* from each site *j* the existence of which it has come to know. To make use of this property, the predicate *ended_i* is introduced:

$$ended_i \equiv ((x, y) \in knownchannels_i \Rightarrow$$
$$(x \in knownsites_i \text{ and } y \in knownsites_i))$$

When this predicate has the value *true*, P_i knows that it has learnt the whole network.

With these elements, the procedure for each site P_i is:

on decision to learn the network
  do
      $participant_i := true$;
      $\forall x \in neighbour_i$: **send** identities $(i,\ neighbours_i)$ **to** P_x
  end do
on reception of identities $(j,\ v)$ **from** P_k
do
  if $\neg participant_i$ **then**
  $participant_i := true$;
      $\forall x \in neighbours_i$: **send** identities$(i,\ neighbours_i)$ **to** P_x
  end if;
  if $j \not\subseteq knownsites_i$ **then**
      $knownsites_i := knownsites_i \cup \{j\}$;
      $knownchannels_i := knownchannels_i \cup \{(j, x): \forall x \in v\}$;
      $\forall x \in neighbours_i - \{k\}$: **send** identities$(j,\ v)$ **to** P_x;
        if $ended_i$ **then** P_i knows all the network **fi**
  end if
end do

As soon as at least one site P_i decides to learn the network it transmits a message *identities(i, v_i)* which will pass through the whole network in a parallel traversal and will also give rise to the transmission of a message *identities(j, v_j)* from each other site P_j and its neighbours, and the predicate *ended_i* will allow it to decide whether or not it knows the entire topology.

In this algorithm the teaching of the network is carried out by all the sites; no single site does it alone.

One *identities(i, v_i)* message at least, and two at the most, are transmitted on a given channel (one in each direction if there are two); if there are *n* sites and *e* channels, the complexity, in terms of number of messages of the algorithm will be, at the worst, $2ne$ messages.

The learning of the network by all the sites is important because, although they are initially dependent on this global knowledge, they can, if they apply the same algorithms, deduce consistent information from it: for example, on routeing, flow, the articulation points of the network, or the determination of the minimum weight spanning tree (which defines in a cycle-free structure a single route between any pair of sites such that the sum of the lengths of the routes is globally optimal).

3.5. Finding the shortest routes

3.5.1. Context of the problem

The problem consists of learning, for each site P_i, its distance from each of the other sites; by distance is meant the shortest distance, with the distance from a site to one of its neighbours being 1. This is the classic problem of the shortest routes through a graph, each edge being valued at 1.

A simple way of solving this problem is to use the previous algorithm; each site P_i, once the network is known, can calculate for each other site P_j the shortest distance (number of edges to traverse between P_i and P_j). It can then use the information obtained to carry out the optimal routeing of messages that will then be transmitted to the other sites.

An algorithm that does not require knowledge of the network by each site will now be introduced. As it concerns a global problem (a site calculates the distance to another site) it is assumed that each site knows the maximum value *maxn* of the number of sites present in the system. (This global information does not invalidate the general assumptions which are made if this value is chosen to be fairly large; cf. Section 3.3.3.)

3.5.2. The algorithm

3.5.2.1. CONTEXT OF THE SITES

A site must store the shortest distance separating it from each other site in the network. If *maxn* is known, each site P_i is provided with the following declaration:

var distance$_i$: array[1 ... masn] of positive or zero integer;

such that the *distance$_i$ [j]* will indicate to P_i its shortest distance from P_j. Since the sites have initially no knowledge of the distances, the table is initialized with $+\infty$. The network is connected and includes n sites; any shorter route between two sites therefore has a distance equal to or less than $n-1$. If n is not known, it is still possible to represent $+\infty$ by *maxn*: any distance along a route will be less than this value. In addition, *distance$_i$ [i]* is initially 0 and will retain this value. If the sites know their neighbours not by the links connecting them but by their identities, it is possible to specify other initial values (which will not be modified by the algorithm).

$$\forall\, j \in neighbours_i <=> distance_i\,[j] = 1$$

The algorithm proposed can benefit from this initial knowledge constituted by the knowledge of the neighbours, but is not dependent on it. The following declarations are made:

var distance$_i$: *array* [1 .. *maxn*] *of* 1 .. *maxn*
  initialized to maxn : $\forall$ *index of j*≠*i and at* 0 *for j=i*;

Knowing the shortest routes is one thing, but knowing how to make use of them is another. To achieve this any site P_i must know which of its neighbours is the best to reach a given site P_j; this information is necessary to achieve optimal routeing of messages. For this each site P_i is given the following table:

var routeing$_i$ table [1 .. *maxn*] *of* 0 .. *maxn initialized to* 0;

Once the shortest values have been calculated, *routeing$_i$* [*j*] will give the 'best neighbour' to P_i to reach P_j, if this site is in the network (this will give *distance$_i$* [*j*] ≠ *maxn*). This array is only used internally by P_i; it is never communicated and could be deleted from the algorithm without affecting the calculation of the shortest route values.

3.5.2.2. PRINCIPLE OF THE ALGORITHM

The principle on which the algorithm is based is simple and similar to an iterative computation. Initially one site (or more) will broadcast its array of distances (that is, its initial knowledge) to its neighbour. When a site receives such information, it updates its context and if this is modified (it learns of distances shorter than those it knew of before) it broadcasts its new table of distances (its new knowledge) to its neighbours. Any site receiving a message will follow this behaviour, and in this way the stages of computation can be conceptually isolated. A site first learns which sites are at a distance of 1, then those at a distance of 2, and so on (in fact this depends on the speed of the messages and the number of sites starting up simultaneously). All the sites are at a finite distance from each other, so *maxn* constitutes a maximum value of the number of such 'stages'.

Once the principle on which the sites cooperate has been established, it remains to state the principle according to which the sites update their contexts on reception of a message. These messages are of a unique type *maxvdist* and carry an array *d*. Consider the site P_i receiving such a message from P_j, its neighbour. *d* represents the value *distance$_j$* at the sending of the message and informs P_i that this value is a new item of information that P_j has not yet transmitted to it. For any potential site P_k ($1 \leq k \leq maxn$), the site P_i will test whether the message received allows it to determine a shorter distance to P_k, in which case it takes into account this new distance and sets a flag which, once all of the distances have been examined, will indicate that it has acquired new information and must transmit it to its neighbours:

> **if** *distance*$_i$ [k] $> 1 + d$ [k]
>   **then** *routing*$_i$ [k] := j;
>   *distance*$_i$ [k] := $1 + d[k]$
>   *modif* := *true*
> **end if**

The variables *distance*$_i$ [k] initialized at *maxn* can only decrease and remain greater than or equal to 1 (with the exception of *distance*$_i$ [i] = 0).

3.5.2.3. THE ALGORITHM

Each site P_i is provided with two tables *distance*$_i$ and *routing*$_i$ to which the boolean *modif* should be added. The text of the algorithm is initially only based on the identity of the channels of communication that connect it to its neighbours. During a preliminary exchange, any site may communicate its identity to its neighbours which then identify it with the corresponding channel. To simplify the expression, it is assumed that the sites know the identities of their neighbours, since this in no way decreases the generality of the solution.

With these elements, the algorithm relative to P_i is as follows:

> **on decision to** *calculate the shortest route values*
>   **do**
> $\forall x \in$ *neighbouring*$_i$: **send** *maxvdist*(*distance*$_i$) **to** P_x
>   **end do**
> **on reception of** *maxvdist* (*d*) *from* P_j
>   **do**
>   *modif* := *false*;
>   **for** *k* **from** 1 **step** 1 **to** *maxn*
>   **do**
>   **if** *distance*$_i$ [k]$>1+d[k]$
>   **then** *routing*$_i$ [k] := j;
>   *distance*$_i$ [k] := $1+d$ [k];
>   *modif* := *true*
>   **end if**
>   **end do**
>   **if** *modif* **then**
> $\forall x \in$ *neighbours*$_i$: **send** *maxvdist*(*distance*$_i$) **to** P_x
>   **end if**
>   **end do**

Note: In an implementation of this distributed algorithm, it would be possible to transmit only the values of the table *distance*$_i$ that have been modified: a list of ordered pairs (k, *distance*$_i$ [k]) would have to be transmitted. The choice of representation is not very important at the abstract level chosen for the expression of this algorithm.

3.5.2.4. TERMINATION

This algorithm does not allow one site to know whether the algorithm has ended or not. Several solutions are possible if this knowledge is required.

One solution would consist of using the network learning algorithm seen above, and providing the sites with an *ad hoc* sequential algorithm.

A more orthodox solution would be to develop an algorithm similar to the one above, but in which the states of computation are strongly synchronized. All the sites start the algorithm (that is, stage 1) simultaneously. This is easy to achieve: the site or sites that have started the computation give rise, by the broadcasting of their first message to their neighbours, to a chain reaction which activates all the sites. Once a site is activated, it becomes a participant and broadcasts its first message to its neighbours. It is during stage p that each site P_i learns the sites that are at a minimal distance p, but this stage would not be started until stage $p-1$ is ended. A site knows that the stage $p-1$ is ended when it has received, during this stage, a message from each of its neighbours. The reception of these messages allows it to pass on to the next stage if it has learnt of a new one or, in the contrary case, to conclude that it will not learn anything else (note that the network is connected); the algorithm is then ended as far as it is concerned. It can inform its neighbours and will discard the messages received. The reader may, as an exercise, give the statement of such an algorithm. He may also extend this algorithm or the previous one so that the sites will also learn the diameter of the network in which they are embedded.

3.6. Resilience to modifications of the network

3.6.1. Types of modification

The network in which the distributed system is implanted can undergo modifications. The importance of the assumptions made concerning the location of initial site knowledge has been explained, in cases where the system is redefined. In certain cases the system must continuously react to modifications to which it is subjected. These are due to breakdowns and their consequences.

There are, firstly, the channels of communication: a channel may break down and will then be out of use; it may be repaired and must then be reintroduced into the system.

Secondly, there is the breakdown of sites. When P_i breaks down, another site is *a priori* unable to tell the difference between the breakdown of P_i and that of the channels connecting it to its neighbours. Since the channels are 'bi-points', the breakdown of the site will be equivalent to that of all the channels connected to it. In the same way, the reintroduction of a site that has been repaired after a breakdown will be equivalent to the reintroduction of the channels to which it is connected.

Methods for making the routeing algorithm given above resilient to breakdowns and channel reinsertions will now be considered.

3.6.2. Breakdowns and channel reinsertions

3.6.2.1. MODIFICATION MESSAGES

It is assumed that the modifications of the network are notified to the sites connected to the channels involved. These notifications are made and arrive in the order of the corresponding modifications for a given channel. It is also assumed, in the interests of simplicity, that the sites know the identities of their neighbours and that any channel is defined by the pair of identities of the two sites it connects. The procedure used here is that of the algorithm described in Section 3.5.2.

Two new types of message are introduced to represent the two modifications: apart from its type, such a message is given its value by the identity of the channel concerned:

breakdown(i, j) and *insertion*(i, j)

These messages are sent to the sites P_i and P_j when the channel (i, j) is subjected to the corresponding event.

The breakdown of a channel can lead to an increase in the values of the shortest routes; its insertion can have the opposite effect. In both cases, at least the sites P_i and P_j are affected by these modifications.

When it receives a message *breakdown*(i, j) the site P_i must take j away from its list of neighbours and transfer the modification of the network to its arrays and to the adjacent sites. The reception of *insertion*(i, j) must produce a similar effect but with the difference that j is here added to the neighbours of P_i.

3.6.2.2. THE INITIAL CONTEXT

In order to resist these modifications, the context of the sites P_i will be modified. Next to *neighbours*$_i$ and *distance*$_i$ a bi-dimensional table is introduced:

var *distfromto*$_i$: **array** $[1 \ .. \ maxn, \ 1 \ .. \ maxn]$ *of* $0 \ .. \ maxn$;

which replaces the array *routing*$_i$. This array, which gives more information, is such that *distfromto*$_i$ $[j, k]$ indicates, if P_j is a neighbour of P_i, the shortest distance between this neighbour P_j and P_k. This array, with *distance*$_i$ allows the best neighbour to be determined for optimal routeing to be achieved; it can yield more information, however, and this will be used to calculate the shortest distances. This array is initialized in the following way:

$\forall j \in neighbouring_i : distfromto_i [j, j] := 0$
$\forall j \notin neighbouring_i, k : distfromto [j, k] : maxn$

As the algorithm must be resistant to the insertion of channels, the initial context may be considered as a network formed of sites unconnected to each other: these are the insertion messages that will therefore make up the network initially:

$neighbours_i = \varnothing$
$distance_i [i] = 0$ *and distance*$_i$ *[j] = maxn for j*$\neq$*i.*

3.6.2.3. SITE BEHAVIOUR

As the system evolves a site can receive an insertion message, a breakdown message or a message to say that one of its neighbours has modified its arrays. Consider first the behaviour of P_i when it receives a message *insertion* (i,j). The first thing this tells it is that P_j is a new neighbour, and therefore it updates *neighbours*$_i$, setting *distance*$_i[j] := 1$ and *distfromto*$_i[j,j] := 0$; and as its context, so far as shortest distances are concerned, has changed it informs its neighbours of this by means of messages of type *majdist*: each such message consists of 3 fields, a distance *dist* and the identifiers *j, k* of the two sites to which it refers. P_i informs P_j also of its shortest distances by means of messages of this same type.

When it receives a message *breakdown*(i,j), P_i modifies its context to show that P_j is no longer one of its neighbours. The effect of this loss is to modify the array *distance*$_i$; using the array *distfromto*$_i$ to find the new values for the routes, P_i calculates the new shortest distances and broadcasts these to its neighbours.

Finally, when P_i receives a message *majdist*$(dist, j, k)$ from a neighbour P_j it updates *distfromto*$_i[j, k]$, calculates its new distance from P_k and if this has changed broadcasts the new value to its neighbours.

Calculation of distances is an important constituent of the algorithm, and as this occurs in several places we give it in abstract form in the procedure below. When this is performed by P_i it gives the 'new' distance from P_i to P_k; if this is different from the 'old' value *distance*$_i[k]$ the updating is performed and the fact signalled by the flag *modif.*

```
procedure calcdist (i, k : 1 .. maxn);
    begin
        if i ≠ k then y := min {distfromto_i [j, k]};
```

$j \in neighbours_i$
if $y < maxn-1$ *then* $d := y+1$ *else* $d:= maxn$
end if
if $d \neq distance_i [k]$ *then* $distance_i [k]:=d;$
$modif := T$
end if
end if
end

The calculation of the distance from P_i to P_k is based on a path through that neighbour of P_i that gives the shortest distance to P_k; it is here that the role of the array $distfromto_i$ becomes clear. If $j \in neighbours_i$, then $distfromto_i [j,j] = 0$, otherwise $= maxn$, so as well as having its main meaning this array can give an expression for the set $neighbours_i$.

3.6.2.4. THE ALGORITHM

The principles just described give the algorithm as follows; the text is written as for the site P_i.

on receiving $insertion(i, j)$
 do $neighbours_i := neighbours_i \cup \{j\};$
 $distance_i [j] := 1;$
 $distfromto_i [j, j] := 0;$
 $\forall x \in neighbours_i - \{j\}:$ *send* $majdist(distance_i[j],i,j)$ *to* P_x
 $k \in 1 .. maxn, k \neq i, j :$ *send* $majdist(distance_i [k], i, k)$ *to* P_j
 end do
on receiving $breakdown(i, j)$
 do $neighbours_i := neighbours_i - \{j\};$
 $\forall x \in 1 .. maxn: distfromto_i [j, x] := maxn;$
 for k *from* 1 *step* 1 *to* $maxn$
 do $modif := F;$
 $calcdist(i, k);$
 if modif then x $neighbours_i :$ *send* $majdist(distance_i [k], i, k)$ *to* $P_x;$
 end if
 end do
on receiving $majdist(distj, j, k)$
 do $distfromto_i [j, k] := distj;$
 $modif := F;$
 $calcdist (i, k);$
 if modif then $\forall x \in neighbours_i :$ *send* $majdist(distance_i [k], i, k)$ *to* $P_x;$
 end if
 end do

3.6.2.5. PROPERTIES OF THE ALGORITHM

This algorithm ensures the maintenance of the routing tables of a network in which channels can break down and then, once repaired or replaced, be reintroduced.

The algorithm ensures the consistency of the arrays; this means that at times when there is no modification under way in the system, and no message in transit or being processed, the routing arrays define the routes that are effectively optimal as such. The arrays then converge towards the description of the optimal routing. If the previous conditions are not all fulfilled, and while there is no modification under way in the network, the system will move to a state in which there are no more messages in transit or being processed. In other words, each modification of the network generates only a finite number of messages for the updating of routing arrays.

One question that remains to be answered concerns the function performed by the algorithm in the maintenance of routing arrays, and the properties it has. When it receives a message m intended for *dest*, the site P_i obeys the following behaviour:

> **on reception of** message $(m, dest)$
> > **do**
> > > **if** $dest = i$ **then** *the message has arrived*
> > > **else if** $distance_i$ $[dest] = maxn$ **then** *delivery impossible*
> > > **else let** $k : |k \in$ *neighbours*$_i$ **and** $(\forall k' \in$ *neighbours*$_i =>$
> > > > $distfromto_i$ $[k, dest] \leq distfromto_i$ $[k', dest])$;
> > > > **send** message$(m, dest)$ **to** P_k
> > > **end if**
> > **end do**

Notifications of breakdowns and insertions of links are received in their order of appearance. Moreover, let us consider that:

—the notification of a breakdown is received after the reception of the messages transmitted before the breakdown;

—the notification of an insertion is received before the messages transmitted after this insertion.

In other words, notifications of breakdowns and insertions are subject to the assumption of non-desequencing of messages on a channel.

Despite these additional stipulations, the breakdown of a channel may give rise to the loss of messages: this is the case when P_i, after having transmitted messages to P_j, learns that the channel has broken down; it is also the case when P_i discards a message with destination P_j just before receiving a message *majdist* informing it of a route to P_j. Messages between two sites may therefore be lost, whether or not these sites belong to the same connected component of the network at a given moment. To avoid such losses, the user of the service must apply adequate control—for example by using sequence numbers associated with the messages transmitted and acknowledgements of reception (end to end protocols). The techniques used in an example in Chapter 1 are quite adequate for this.

3.7. A general principle

In a certain number of algorithms introduced in this chapter (and in Section 2.4, on resilience to breakdowns in the clock problem), a single principle is used. A site, whether as a result of a local decision, or on the reception of a message from one of its neighbours, acquires new information that modifies its knowledge; it then broadcasts this new knowledge to its neighbours; if, on the other hand, it learns nothing new when it receives a message, it does not engage in any communication. This 'principle of extinction-broadcast' is general; many algorithms use it in a variety of forms. When, for example, the messages are dated or numbered, the control information constituted by the date or the number indicates whether the message received is out of date (extinction) or if it must be taken into account. The principle has been used in this chapter in the network learning algorithm (when P_i receives a message (*identities(j, v)*)) and in the two algorithms for computation of routing (with and without modification of the network).

3.8. References

Network traversal has been studied by many authors and in many different ways.

Cheung (1983) uses traversals to calculate the maximum flow through a network; Segall (1982) does the same. Chang (1983), adopting the name 'echo algorithm', implements a parallel traversal which he uses to calculate biconnected components; for the same calculation Awerbuch (1985) proposes a depth-first traversal. Dijkstra and Scholten (1980) use a tree-type traversal as the basis for an algorithm for detection and termination. Helary, Maddi and Raynal (1987) also propose such traversals, combined with knowledge of the identity of neighbours, to achieve elections in an arbitrary network.

For more general concepts, the reader will find in Gondran & Minoux (1979), Even (1979) and Aho (1974) sequential algorithms on graphs used locally by network sites; in Tanenbaum (1981) and Pujolle *et al.* (1985) there are more comprehensive networks concepts.

Distributed algorithms determining a minimum weight spanning tree for any network have been particularly thoroughly studied. These algorithms are very important because of the global optimization they achieve. Gallager *et al.* (1983) and Lavallee and Roucairol (1986) suggest a solution for non-oriented graphs, while Humblet (1983) is concerned with oriented graphs. Many other aspects of networks (graphs) have been studied in a distributed context: the calculation of centres and medians in networks (Korach *et al.* [1984]), the calculation of shortest routes with the edges weighted at 1 (Segall [1983]) or arbitrary values

(Chandy [1982]) and the calculation of connected components (Misra and Chandy [1982]), etc.

The problems of learning the network have many solutions; Segall (1983) studies several of these. MacCurley and Schneider (1986) propose and prove a learning algorithm for a highly connected network in the case where the channels are unidirectional. Bouge (1985) considers the same problem in the case of networks expressed in the CSP formalism.

The problem of resilience to channel breakdowns has also been considered. The algorithm presented has been proved correct by Lamport (1982b). This algorithm is a variation of the one proposed by Tajibnapis (1977). Merlin and Segall (1979) also propose an adaptive routing algorithm that periodically recomputes the best routes as a function of the load on the channels, and channel breakdowns and reinsertions. This algorithm also maintains, for any site, the tree-structure of the shortest routes from this site to all the others; Merlin (1979b) gives a proof of its correct functioning.

Chapter 4

Determination of a global state

4.1. The global state problem

4.1.1. Introduction

In the previous chapter, a number of basic algorithms were introduced which, when applied in a suitable software layer or combined with other algorithms, allow problems linked to the distributed system structure and its use to be solved. With these problems of addressing and knowledge solved, the implementation of the system may require from the sites a 'correct' view of its state; the detection of deadlocks, for example, relies on the knowledge of a state of the system which indicates that each site in a set of sites S is halted while waiting for a message from a site in S, and that between the sites of S there are no messages in transit. A state of this kind is global: it involves a certain number of sites and communication links in the system, possibly all of them.

The solution of certain problems linked to the control that must be ensured by the system will therefore require knowledge of the global state of the system by the site or sites charged with control.

The first two chapters contained examples of applications of control (mutual dependence of a producer and consumer, and avoidance of drifts between logical clocks), from which the following conclusions can be drawn:

1) at a given moment a site has only an approximate knowledge of the state of the other sites;

2) the events that take place in a certain order at one site can be perceived in a different way at other sites.

To illustrate these two points, consider the example of the logical clocks in Chapter 2, in which three sites P_1, P_2 and P_3 are considered, with their respective clocks h_1, h_2 and h_3. Each clock h_i, from the point of view considered here, describes the state of site P_i. Taking into account the losses and delays in transmission of messages, at a given moment the state of P_2 and its perception by P_1 and P_3 might be the following (point 1 above):

—state of P_2 : $h_2 = x$
—perception of this state by P_1 : $last_1 [2] = y$ with $y \leq x$
—perception of this state by P_3 : $last_3 [2] = z$ with $z \neq y$ and $z \leq x$

Now consider two events e_1 and e_2 located at P_2: transmission of two successive and different values of h_2. The events relating to the corresponding reception can be perceived in the following way (point 2 above):

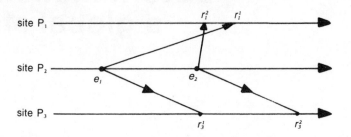

This absence of a global state that could be detected *a priori* by a site is a defect as regards the solution of certain problems. It is therefore important to construct a distributed algorithm that will allow one or more sites in the system to obtain a correct global state. This chapter concerns the design of such an algorithm. It is a learning algorithm, but unlike those in the previous chapters, there are severe constraints on the rules governing the obtaining of information (using messages) such that the composition of the local states obtained gives a consistent global state as its result.

4.1.2. Context and definitions

There are several ways of describing a distributed system, according to the orientation given by the description: towards the actions or operations executed by the sites or towards the states through which the system passes. In the first case, the expression used is the *space–time* view of the system; in the second, the *interleaving* view. In order to clarify these concepts a certain number of assumptions and definitions will be made that will run through the chapter.

4.1.2.1. ASSUMPTIONS

The system considered is made up of a set of sites: each site is the location at which a sequential process is executed (site and process will be used interchangeably in the following, but this does not lead to confusion). The sites are connected by a network of unidirectional links also called channels which are reliable: no loss, duplication or desequencing of messages occurs, and delivery is in an arbitrary but finite time. In addition, the graph associated with the network is assumed to be strongly connected.

4.1.2.2. OPERATIONAL VIEW

An event is the sending or reception of a message, or else it is internal to a process (time out on a guard delay, for example). An event is located in a given process and causes this process to pass from one internal state to another. Since a message cannot be received before being transmitted, the event associated with its reception is always later than that associated with its transmission; this property defines a relationship called the *causality relationship*. The events internal to a given site are totally ordered (issued from a sequential process); the relationship of causality creates a partial order on the set of events occurring at all the sites, which defines the space–time view of the system. This is an essentially operational view of a distributed system.

4.1.2.3. OBSERVATIONAL VIEW

When it is to be established whether or not a system has a given property, that system must be observed, and this is generally an observation of the global state. This state is formed by the union of the local states of all the processes that form the system and the states of all the channels (links) of the network on which the system is constructed.

The state of a process is defined by that of its variables (called state variables or state vector); that of a channel is defined by the sequence of messages that it is currently transmitting. A process changes state during an event local to this process; this event can also change the state of a channel to which the process is connected e. g. by addition of a message to or by suppression of a message in transit. In considering the global state of the system, it can be seen that an event produces another global state, identical to the first except in the process and channel affected by the event. The progress of the system can therefore be viewed as the succession of events which move the system from its initial state, with processes in their initial states and channels empty to its current state, through a succession of global states. It is this view that is known as *interleaving*; it is a purely abstract view—several events can occur 'simultaneously' in several processes, with the system not in reality progressing sequentially from one event to the next.

4.1.3. The global state problem

4.1.3.1. GENERAL CHARACTERISTICS

Allowing a process P_i to obtain a global state of the system is of interest in more than one respect. Firstly it is possible to apply a centralized algorithm to this state, which will analyse it and give the desired information to the process P_i. This is, in fact, the only possibility when a centralized algorithm is known for a given problem, but not a distributed one. Secondly, obtaining this global state is of interest in the detection of stable properties. These are properties which, once they occur, continue to hold (for example, deadlock, termination, etc). The control process charged with detecting such a property computes a global state, tests the property and starts again if it does not hold.

Providing an algorithm for the calculation of a global state is a basic problem in distributed systems. An algorithm of this type would allow an operational (space–

time) view of a distributed system to be used to obtain an observational view. Starting with the operations of message transmission and reception that constitute the activity of a distributed system, the algorithm calculates a global state in which all operational characteristics have disappeared. By analogy, it can be said that such an algorithm allows a change in the 'co-ordinates system' within which the system is perceived.

4.1.3.2. DIFFICULTIES WITH THE ALGORITHM

Before introducing the algorithm, it is important to be aware of the complex nature of the solution. A process wishing to obtain a global state must not only obtain the local state of each of the other processes and the state of each of the channels, but must do so in a specific way. These states are linked to each other and their combination must give a consistent global state—that is, one which could have been observed if the events of which the processes are the locations had occurred sequentially.

One difficulty lies in the perception of the state of the channels at a given moment: the sequence of messages sent but not yet received. A simplifying assumption could be to consider the channels always to be empty—that is, that the time of transit is zero, or that the exchanges obeys the rendezvous protocol. The first case is unrealistic and the second simply transfers the problem to the level of implementing the rendezvous. Also, even on such an assumption, a process connected to all the others will have to ask them for their respective states in order to construct a global state and its requests may reach other processes so as to make the combination of the local states obtained consistent. Consider, for example, the states sent to the inquirer P_k by P_i and P_j which have received the requests respectively before and after a communication between themselves.

P_i : communicates with P_j
  receives the request from P_k and sends its state
P_j : receives the request from P_k and sends its state
  communicates with P_i

The state of P_i recorded by P_k indicates a communication with P_j which will not be found in the state of the latter, yet the channels are empty. (In this example the rendezvous applies only to the calculation, and not to the control of the calculation.)

The algorithm for obtaining a global state of the distributed system is therefore not simple: the object being observed may be modified during observation but the observation must not alter it in any way. An algorithm of this type will be derived from a centralized solution that can gradually be distributed. The method for derivation proposed is general and can therefore be used to obtain other algorithms from centralized ones. Apart from the algorithm developed, the aim of this chapter is also to introduce a methodology for the design of distributed algorithms.

4.2. Derivation of an algorithm

4.2.1. A centralized solution

4.2.1.1. PRINCIPLE

Consider the problem of calculating a consistent global state in the following abstract framework:

1) there is a single global temporal reference (in other words, all the processes have access to a common clock);

2) it is assumed that all the processes know *a priori* the date d at which they must store their respective local states (d is relative to the global clock).

These two additional assumptions define a centralized system context without common memory and the knowledge that the processes must have *a priori*. Under these assumptions a simple but not distributed algorithm will be developed. The method consists of weakening these constraints while retaining the result, so that the new assumptions take account of a distributed system.

If each process associates the date of transmission with any message transmitted, hypothesis will allow it to order the messages transmitted before d—known to them all, by hypothesis 2—and received after d: these are the messages in transit at date d.

4.2.1.2. THE ALGORITHM

The global clock gc is accessible to all the processes, which are connected by unidirectional channels, without loss, desequencing or duplication.

The context of P_i is as follows:

var cin$_i$: array $[1 .. k]$ of input channels;
    cout$_i$: array $[1 .. l]$ of output channels;
    chanstate$_i$: array $[1 .. k]$ of message sequence initialized to $\varnothing$;
    received$_i$: array $[1 .. k]$ of boolean initialized to false;

The variable *chanstate$_i$* $[x]$ will allow P_i to record the state of its input channel $cin_i[x]$ at date d: for this P_i adds any message transmitted at or before d and received on $cin_i[x]$ at a date later than d. A message of this type would be found in transit at date d, to which the calculated global state corresponds. When P_i has received on each of its input channels a message transmitted after d it knows the state of all its input channels at date d and can then send the state of its input channels and its local state ls_i which it recorded at date d to a particular process charged with collecting the global state. ($\oplus$ designates concatenation of messages.)

on transmission of m on cout$_i$ $[x]$
    do
        dm := read(gc);

```
        send(m, dm) on cout_i [x]
    end do
at date d
    do
        record local state ls_i
    end do
on reception of (m, h) on cin_i [x]
    do
        if d < read(gc) and h ≤ d
        then chanstate_i [x] := chanstate_i [x] ⊕ (m, h)
        else if h > d
        then received_i [x] := true
        end if
        if (∀ x ∈ 1 .. k : received [x])
        then send (ls, ∀x : chanstate_i [x]) to collector process
        end if
    end do
```

4.2.1.3. A CONTROL MESSAGE: THE MARKER

The algorithm is simple: each process stores its local state and that of each of its input channels at that date; this involves the messages transmitted before or at d and received after d.

This centralized algorithm does, however, have a disadvantage: the delay in the transit of messages on the channels is finite but arbitrary, so a process must wait to receive a message transmitted after d on each of its input channels to learn the state at instant d (the messages do not overtake each other).

This leads to difficulty if a process sends its last message on a channel before date d, which can be eliminated by the systematic sending of a special message at instant d by all the processes on all the channels: such a control message, called a marker, plays a similar role to that of the sentinal values in sequential applications, to avoid the loop index exceeding the limits of an array. The use of such a message will be explained in the final version of the algorithm.

4.2.2. Local logical clocks: use of a virtual global time

Hypothesis 1) above, on the existence of a global clock used by all the processes to date their messages, can now be eliminated. The properties of the global clock used by the algorithm are as follows:

p1) the delays in transit are finite, arbitrary and non-zero, so a message is always received after its transmission;

p2) in a given process, two message transmissions at different moments have different dates associated with them;

p3) the progression of the global clock is such that the d date is reached.

What is needed is the introduction of a mechanism that takes account of these properties of consistent global clock progression. To this end, logical clocks are introduced, represented by monotonic counters initialized to 0. Each P_i is therefore provided with the following control variable:

var h_i monotonically increasing integer initialized to 0

Between two internal events, P_i increases its clock, which takes account of p2) and p3). Taking account of p1) means that when P_i receives a message dated h, its local clock indicates a time later than h; this is achieved by the systematic updating of h_i during reception of a message (m, h):

$$h_i := max(h, h_i) + 1$$

Such a mechanism allows all the events of the system to be dated in such a way that all the events of a process are ordered by their (logical) dates of occurrence, and the (logical) dates associated with the communication of a message are such that the date of despatch is less than the date of reception. The activity of the system can therefore be modelled by a partial ordering of the events in it. This is the operational or space–time view.

4.2.3. An initial algorithm

Starting with the algorithm based on a global clock accessible to all processes, it is possible to derive a distributed algorithm simply by replacing this global clock by local logical clocks and the mechanism for clock progression and updating described above.

When a process wishes to bring about the calculation of a global state, it need only advance h_i to the value d; the messages it then sends to its neighbours will have dates greater than d, leading to the storage of the local states of processes and channels relative to this date. The global state obtained is therefore relative to the abstract and global date d. In addition, it does not depend on the number of processes that have taken the initiative for its calculation by causing their clocks to move forward to d.

As indicated, this method of deriving a distributed algorithm from an algorithm based on a global clock is general: it provides an implementation of the global clock which is simply a conceptual abstraction making the design of the algorithm easier.

4.3. The Chandy and Lamport algorithm

4.3.1. Elimination of logical clocks

We have seen that the previous algorithm assumes that each process sends messages after date d so that their reception leads to the state of the corresponding channel being stored. The systematic sending at date d of marker-type control messages has eliminated this constraint and the problem of processes that send no more messages after d. On a given channel there are messages transmitted before d and others transmitted after—that is, before and after the marker. But the algorithm is based only on the fact that a message is transmitted before or after d—that is, before or after the marker: the logical clocks can therefore be eliminated and only the markers used, with the occurrence of date d for a process corresponding to its first transmission, or reception of a marker.

The algorithm therefore appears in the following form: when P_i decides to initiate a global state calculation, it stores its local present state and sends markers on its output channels. When P_j receives a marker on an input channel $cin_j [x]$, then, if it has not recorded its own state it proceeds as if P_i; if it has recorded its state it takes the state of the channel $cin_j [x]$ to be the sequence of messages received after the recording of its state and before the marker.

4.3.2. The algorithm

Each process P_i is equipped with the same procedure as before, to which is added the boolean:

*var storestate$_i$: **boolean initialized to false**;*

which specifies whether P_i participates in the calculation—that is, whether it has recorded its state.

> **on decision** *to initiate calculation*
>   **possible only if** *storestate$_i$ = false;*
>   **do**
>   *record local state ls$_i$; storestate$_i$:= true;*
> $\forall y \in 1 .. l$: ***send****(marker)* **on** *cout$_i$[x];*
> $\forall y \in 1 .. k$: *chanstate$_i$ [y] := \varnothing;*
>   **end do**
> **on reception of** *m* **on** *cin$_i$ [x]*
>   **do**
>   ***case*** *m = marker **and*** ⌐ *storestate$_i$→record local state ls$_i$;*
> $\forall y \in 1 .. l;$
>   ***send****(marker)* **on** *cout$_i$[y];*
> $\forall y \in 1 .. k$: *chanstate$_i$ [x] := \varnothing;*
>   *storestate$_i$:= true;*
>   *received$_i$ [x] := true*
>   *m = marker **and** storestate$_i$* → *received$_i$ [x] := **true**;*
>   **if** $\forall x$: *received$_i$ [x]* **then**
>   ***send****(ls$_i$, U chanstate$_i$ [x])*
> $x \in 1 .. k$
>   *to collector process*
>   **end if**

$$m \neq marker \text{ and } storestate_i$$
$$\text{and } \neg received_i [x] \rightarrow chanstate_i [x] := chanstate_i [x]$$
$$\oplus m;$$

case
end do

4.3.3. Termination

Several processes can 'simultaneously' record their state and consequently send markers on the output channels. If a process does not have this capability and has no input channels, and therefore receives no markers, it cannot record its state, and will not transmit either its local state or any markers on its output channels: the algorithm for calculating a global state will therefore not terminate.

To avoid these problems it will be assumed that the network on which the system is constructed is modelled on a highly connected oriented graph. This connection ensures that when a process initiates a calculation, all processes will be reached and all channels traversed by the markers. In fact, once P_i initiates a calculation all the P_j such that there exists a route from P_i to P_j will participate in the calculation; note that the transmission delays are finite.

4.3.4. The Chandy and Lamport formulation

The above algorithm can be reformulated as two rules defining the management of the markers by P_i. This is the original formulation of the algorithm.
Rule 1: sending markers.
  P_i sends a marker on each of its output channels after having recorded its state and before transmitting other messages on these channels.
Rule 2: reception of a marker on $cin_i [x]$
  If P_i has not recorded its state
  then it stores it, sends markers on its output channels and records the empty sequence for the state of $cin_i [x]$
  else it records the state of $cin_i [x]$ as the sequence of messages received on that channel between the recording of its state and the arrival of this marker.

Note: In the case where a single process initiates the calculation, the markers carry out a 'highly structured' parallel network traversal, the associated graph of which is highly connected, cf. Chapter 3. A marker travels along each arc. In the manner of a syntactic analysis on which are grafted semantic and code-generation actions, the arrival of a message gives rise to 'table updating' ($chanstate_i [x]$) and 'generation' actions (sending the recorded states) when the completed tables can be used.

4.3.5. Generalization

Several global states can be calculated: to distinguish between them, they can be assigned distinct identities—for example, by associating them with the identity of the process initiator or with a colour. In the first case, the markers with the same number relate to a calculation initiated by a single process, while in the second the markers of the same colour relate to a global state calculation initiated by one or more processes. Several global state calculations can then coexist, with each corresponding to a given temporal cut-off (cf. Section 4.3.6.).

It is possible to weaken these restrictions on the channels by introducing suitable mechanisms. The desequencing of messages, for example, can be tolerated: acknowledgements of the reception of each message (by the underlying calculation or by the control) must be introduced, which will make the sending of any message subject to the reception of an acknowledgement of the message previously transmitted on the same channel. The original sequencing is therefore reconstructed. In the case of losses or duplications, sequence numbers must be introduced (cf. Chapter 1).

4.3.6. Characteristics of the global states obtained

In order to use the global state obtained, it is important to specify what it represents. The systematic use of markers by the algorithm indicates that the recorded local states and channel states relate to events which have taken place before a certain date or between two dates.

4.3.6.1. GLOBAL STATE AND EVENT SET

To understand this idea, consider the concept of temporal cutoff, which allows the events produced by the system to be divided into two subsets, respectively relating to the periods before and after the cutoff.

A cutoff C is a set of events such that if the event $g \in C$ and f is an event that precedes g (the two events have therefore been produced by the same process in that order, or have been produced by distinct processes and f is or precedes the transmission of a message the reception of which is or precedes g), then $f \in C$.

It is easy to prove that the cutoffs of a system are not necessarily totally ordered but that the union and intersection of cutoffs are cutoffs. The graph of the 'precedence' relation over the set of cutoffs of a process is determined by the inclusion relation.

The cutoff concept therefore reveals the relationships between events: precedence within a process and causality between processes. It is therefore possible to associate a global state $S(C)$ of the system with a cutoff C in the following way:

—for each process P_i, the state of P_i in $S(C)$ is the state resulting from the last event of P_i in C if it exists, otherwise its initial state.

—for each channel C_j, the state of C_j in $S(C)$ is the sequence of the messages sent by the events belonging to C and received by events not belonging to C.

Consider, for example, the following sequence of events involving three processes: an event is shown by a cross, the arrows linking two events belonging to distinct processes show the transmission and reception of messages. For a given process, the events succeed each other from left to right.

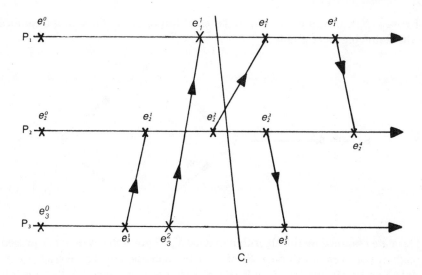

Only the events involving communications have been indicated. Between two events in a process a finite number of internal events can be added. $C_1 = \{e_1^0, e_1^1, e_2^0,$ $e_2^1, e_2^2, e_3^0, e_3^1, e_3^2\}$ is a cutoff whereas $C_1 - \{e_3^2\}$ is not: this set contains an event associated with a reception and does not contain the event associated with the corresponding transmission. A certain number of cutoffs can therefore be constructed: each one represents the history of the system that has led to the corresponding global state.

A global state produced by the algorithm therefore corresponds to a history of the system, resulting from the occurrence of events in the corresponding cutoff.

4.3.6.2. PROPERTIES OF THE STATE OBTAINED

When the global state calculation algorithm is started, the system is in state S_0 (which is not known). When it ends it will be in the global state S_1 (which is not known *a priori*). The calculated global state S is such that:

—from S_0 the system can reach S, and
—from S the system can reach S_1.

Between S_0 and S_1 therefore there exist trajectories in state space and at least one of them passes through S, although nothing indicates that the system has 'really' passed through this state. To make this result clearer, consider the previous example and the following global states:

S_0 corresponds to the cutoff $C_1 - \{e_2^2\}$
S_2 corresponds to C_1
S_3 corresponds to $C_1 \cup \{e_1^2\}$
S_4 corresponds to $C_1 \cup \{e_2^3\}$
S_1 corresponds to $C_1 \cup \{e_1^2, e_2^3\}$

Between S_0 and S_1 the system can therefore follow one of the routes described by the following state graph:

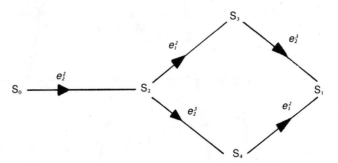

The state obtained by the algorithm may be S_4 whereas the system has passed through S_3 (or even not through S_3 or S_4 if the reception of the message by P_1 (event e_1^2) and the transmission by P_2 (event e_2^3) have taken place simultaneously).

It is possible to show that there is a permutation of independent events, in the sequence of events relative to the execution of the system, gives rise to a sequence that makes the system pass through the state obtained.

4.4. Two uses of the global states

4.4.1. Detection of stable properties

A stable property of a system is one which, when it holds, will continue to do so during subsequent evolution of the system. Examples of stable properties have already been cited: termination and deadlock are particular representatives. More formally, the predicate P is a stable property of the system if:

$\forall$ global state S' attainable from $S : P(S) => P(S')$.

4.4.1.1. GENERAL DETECTION ALGORITHM

A detection algorithm for a stable property P can naturally be based on the calculation of a global state. It takes as input the property P and gives as result a boolean *stable* such that with S_{init} and S_{end} as the global states of the system at the start and end of the detection algorithm, the following is found:

$(P(S_{init}) => stable)$ *and* $(stable => P(S_{end}))$

A specific process can carry out the detection in the following way:

> *stable* := *false*;
> **while** $\neg$ *stable* **do**
> > *initiate calculation of a global state*;
> > *state S is obtained*;
> > *stable* := $P(S)$;
> > *output*(*stable*)
> **end do**

4.4.1.2. INTERPRETATION OF THE RESULT

When the result *stable* = *false* is obtained, the algorithm continues to obtain global states and to test them for P. From *stable* = *false* it can be deduced that $P(S_{init})$ is also false, but nothing can be determined about $P(S_{end})$.

When *stable* = *true*, stability has been detected and it is deduced that $P(S_{end}) =$ *true*; on the other hand nothing can be concluded on $P(S_{init})$.

The stability detection algorithm is correct because the global state calculation ensures that any state S obtained can be reached from S_{init} and that S_{end} can be reached from S; since the property P is one of stability, according to the stability definition $P(S) => P(S_{end})$.

Consider, for example, the property of termination, which is a stable property. The predicate *Term* on an obtained global state is:

—$\forall$ processes $P_i : state(P_i) = passive$
—$\forall$ channels $c_j : state(c_j) = empty$

Thus according to the value of *Term*(S):

Term(S) $=>$ *Term*(S_{end}) and
$\neg$ *Term*(S) $=> \neg$*Term*(S_{init})

4.4.1.3. SPECIFIC ALGORITHMS

The detection algorithm introduced is general: it does not depend on the stable property P. If the terminology of abstract data type is adopted this algorithm can be called *generic*, which means that it admits as parameter any predicate P on condition that it expresses a stable property.

When the stability property has been defined it is possible to conceive specific detection algorithms; this is the case for the detection of deadlocking and of termination, for which there are many algorithms. The existence of this general algorithm is significant for more than one reason. It allows new specific algorithms to be derived when P is known, but above all it provides a unique conceptual framework that assists in the understanding of problems, since it facilitates the identification of their common points and allows the fundamental elements to be extracted.

4.4.2. Recovery after breakdown

4.4.2.1. THE RESTART PRINCIPLE

Obtaining a global state for a system is an essential element in resilience to breakdowns: it allows a consistent state to be restored from which the system is recovered. The Chandy and Lamport algorithm can therefore be used periodically to obtain restart points. It must be used in conjunction with a distributed restart algorithm allowing the consistent recovery of the last state stored.

If S is the last global state saved it can be distributed: each process P_i saves locally its own state ls_i and that of its input channels cin_i.

The principle of restarting is simple. As before, consider the centralized context defined by a single global time and let all processes know the date dr at which restart will take place (this is easy to achieve with the broadcast of a control message *restartat*(dr) to all the processes before recovery). There are three elements involved in the principle:

—dating: all the messages are dated using a global clock. This will allow the time of their transmission to be determined with respect to the date ds of calculation of the saved global state S.

—purge and recovery at instant dr each process must purge its input channels of messages transmitted after ds and restore its state and that of its input channels with the corresponding states saved at ds and stored in S.

4.4.2.2. A RECOVERY ALGORITHM

By applying, as in Section 4.2, the principle of global virtual time with local logic clocks and a mechanism for resynchronization at the reception of messages, a distributed algorithm for recovery after breakdowns can be obtained.

All the processes P_i equipped with a logical clock h_i obey the following behaviour pattern, in which ls_i and *chanstate*$_i$ $[x]$, $1 \leq x \leq k$ represent the local state of P_i and the states of its input channels saved in S at date ds. All messages transmitted by the processes are marked with their logical date of transmission (cf. Section 4.2.2.)

The purging of messages transmitted between the dates *ds* and *dr* is achieved by examining the date associated with the message when it arrives. The propagation of the recovery through the system takes place with the transmission of a message with date greater than or equal to the restart date *dr* (a control message of a suitable type may also be used).

> **on the decision to initiate the restart**
> **do**
> $$h_i := dr$$
> **end do**
> **when** $h_i \geq dr$
> **do**
> > **if** restart$_i$ = false **then**
> > **send** (m, h_i) **on** cout$_i$ [x], $\forall x \in 1 .. l$;
> > *restore* ls$_i$;
> > > *restore* cin$_i$ [x] := chanstate$_i$ [x], $\forall x \in 1 .. k$;
> > > restart$_i$:= true
> > **end if**
> **end do**
> **on reception of** (m, h) **on** cin$_i$ [x]
> **do**
> > **if** restart$_i$ **and** ds < h < dr **then** reject message
> > > > **else** accept it
> > **end if**
> **end do**

The restart will spread through the whole system (the network is strongly connected; cf. the traversal algorithms). The restored state S relates to the date *ds*, and the messages transmitted between *ds* and *dr* are purged. As shown, the messages restored in the input channels are received twice: the first time they are acted on by P_i, which also stores them in *chanstate$_i$* [x]; after restoration they are again received and acted on, since P_i is then re-started from the control point represented by *ls$_i$*.

It will be noted that the test *ds* < *h* is of no use at the reception of a message. The messages do not overtake each other, those of them that are labelled with a date lower than *ds* will have been received by P_i and absorbed in the calculation that takes in *ls$_i$*. The test has been left in the text merely as a comment to clarify the meaning.

4.4.2.3. ELIMINATING THE CLOCKS

As in the algorithm for calculation of a global state, the property used here to show whether a message should be rejected is its position before or after *dr*. Markers can be used to express this. When associated with the restart, they allow it to be propagated systematically without the use of the underlying calculation messages. Control variables must therefore be added to the context of the processes to indicate for each input channel whether, once the restart has taken place, a marker has been received or not. In the latter case the messages received are rejected, since the messages reintroduced through each channel (*chanstate$_i$* [*x*]) are after the marker that arrives on this channel.

4.5. An example

4.5.1. A very simple system

Consider a system made up of two processes P and Q and two channels $c1$ and $c2$ which connect them in each direction.

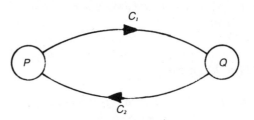

These two processes can send each other messages. Their individual behaviour is the same and can be modelled by the following finite-state automaton.

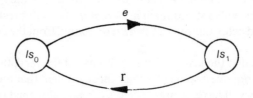

in which: ls_0 is the initial state, ls_1 an intermediate state
e is the transition ***send m on*** c_i
r is the transition ***receive m' on*** c_j
The behaviour of each process is therefore perfectly deterministic. It is a sequence of events in which transmissions and receptions alternate. Initially the channels are empty.

4.5.2. Graph of the global states

In order to study this system a graph describing its possible behaviour is constructed. A global state is made up of the local state of each process and of each of the channels. It can therefore be represented by the following 4-tuple:

$<$ local state of P, state of $c1$, state of $c2$, *state of Q* $>$

(In order to clarify the notation the name of the process concerned is added to the local states and to the transitions. The state of the channel is the sequence of messages in transit described from left to right.)

Starting with the initial state S_0 defined by $<lsp_0, \varnothing, \varnothing, lsq_0>$, the following graph, modelling the behaviour of the system, is obtained:

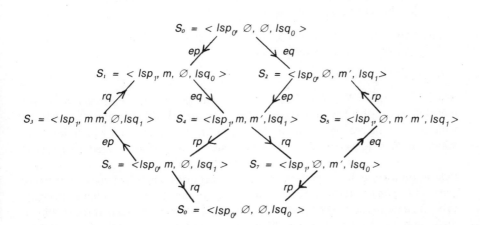

Only the number of messages present is considered in the state of the channel. The finiteness of the graph shows that the size of the channels is limited to 2. (If the message values in the channel states had been considered, the graph could not have been finite, but each channel state would have had between 0 and 2 messages.)

Although the processes, taken in isolation, have a deterministic behaviour, the graph of the global states (which is a deterministic finite-state automaton) describes a non-deterministic behaviour of the distributed system. There are states in the system in which several transitions are *a priori* possible; this is, for example, the case with the state S_4 from which the two transitions rp and rq are possible, with the resulting states being distinct.

4.5.3. A global state calculation

Consider the following specific execution of the system in which global states and the corresponding transitions appear:

$$S_0 \xrightarrow{\quad ep \quad} S_1 \xrightarrow{\quad eq \quad} S_4 \xrightarrow{\quad rp \quad} S_6 \longrightarrow \ldots$$

The calculation of a global state is initiated by the process P when the system is in state S_0. The following events then occur:

el : P stores its local state lsp_0 and sends *marker* on channel $c1$

e2 : P sends the message m on $c1$ (transition ep)

e3 : q sends the message m' on $c2$ (transition eq)

e4 : P receives m' on $c2$ (transition rp) and stores it in *chanstate* $[c2]$

e5 : Q receives the *marker*, records its state lsq_1 and that of the channel $c1$ as equal to $\varnothing$, then transmits the *marker* on its output channel $c2$

e6 : P receives the *marker* on $c2$ and then considers the state of $c2$ as equal to the sequence containing m'.

The state calculation ends at S_6 and the combination of the states stored by P and Q give the following global state:

$$<lsp_0, \varnothing, m', lsq_1>$$

which is the state called S_2 in the state graph. The global state obtained is a consistent state accessible from the state S_0 (in which the algorithm is started) and from which S_6 can be reached (state at the end of the algorithm), but which does not appear in the sequence of states in the particular sequence considered. In the state S_0 the system can traverse one or other of the transitions ep and eq, which are independent there. It is therefore possible to permute them. The two sequences ep eq rp and eq ep rp produce the same result: they make the system change from S_0 to S_6. The global state obtained, S_2, is therefore a consistent state through which the system has been able to pass between the start and the end of the algorithm, respectively identified by S_0 and S_6. If a breakdown occurs at S_6, for example, any of the states S_0, S_1, S_2, S_4 or S_6 can constitute a consistent recovery point.

This example shows that an arbitrary combination of local states and channel states is not necessarily a global consistent state—for example, $<lsp_1, \varnothing, \varnothing, lsq_1>$; the algorithm cannot, of course, produce such a state as its result.

4.5.4. Space–time view

The sequence of system states can be represented by the help of partial ordering of the events occurring in it. The following partial order (temporal precedence relationship) is obtained, with the events e_1, e_5 and e_6 relative to the state calculation and e_2, e_3 and e_4 to the underlying calculation:

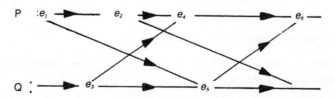

The global state S_2 obtained of course takes account of the execution proper of the system observed and not of the events involved in its observation (e_1, e_5 and e_6 therefore do not appear):

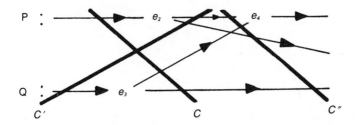

It corresponds to the cutoff C in which P is in the initial state, Q has sent a message, the channel $c1$ is empty and $c2$ contains a message. Different results could be obtained whenever the algorithm is executed—S_1 or S_4, for example, which correspond respectively to the cutoffs C' and C''.

4.6. References

The global state problem is central to the understanding, design and implementation of distributed systems, and is discussed in many publications (Le Lann [1977], Lamport [1984b]). The first solution given is that used in this chapter, devised by Chandy and Lamport (1985), it has been taken up by a number of authors. The introduction given here was largely inspired by Morgan (1985) and Chandy and Lamport (1985).

The different perceptions of a distributed system have been studied in particular by Lamport (1978) and Lamport and Schneider (1985). Logic clocks and their systematic resynchronization are introduced in Lamport (1978), where for example they are used to form a distributed mutual exclusion algorithm.

The algorithm for recovery after breakdown given in this chapter is taken from Morgan (1985), who also introduces the method for deriving a distributed algorithm from an algorithm based on a global clock. This method is also given by Apt and Richier (1985) who use it to obtain termination detection algorithms.

In the context of distributed databases, Traiger *et al.* (1982) use a similar principle for the adaptation of algorithms controlling competition in a centralized resource.

Chandy and Misra (1985) suggest another general schema for the detection of stable properties and derive several specific algorithms from it. Many specific algorithms have been proposed for the detection of stable properties: for example, Chandy *et al.* (1983), Helary *et al.* (1986b), Natarajan (1986) for deadlock and Dijkstra and Scholten (1980), Dijkstra *et al.* (1983), Rana (1983), Apt (1986) for termination. The interested reader will find a synthetic presentation of several of these algorithms in Raynal (1985).

Part 3

Tools, methods and algorithms

The understanding of distributed applications and systems requires knowledge both of schemas for distributed processing (which can define the underlying distributed kernel) and of suitable algorithmic tools and their properties. This third part concerns these two aspects.

The schemas are considered with the distribution of global synchronization constraints in mind; the protocols necessary for their implementation are examples of distributed control.

Starting with the idea of the toolbox, elements that make the solution of problems easier are introduced and analysed. Their use and the effects of their properties are illustrated with algorithms for distributed control and computation.

Chapter 5

Distributing a global synchronization constraint

5.1. Introduction

5.1.1. Synchronization problems

The processes of a distributed application or system are generally called on to synchronize themselves. The aim of this is to ensure a correct allocation of the virtual or physical resources for which they compete, or to guarantee proper cooperation towards their common aim. The synchronization is embodied in expressions that are generally said to define the constraints or conditions requiring the processes to respect certain rules. As a result, when the activity of a process would lead it to break one of these rules, the obligation to respect the constraints prevents it. The activity will then be halted until the corresponding synchronization expression authorizes it. Blocking and signalling primitives are provided for this reason; cf. the monitor introduced in Section 1.2.

These problems are independent of the centralized or distributed nature of the system, and are linked only to the existence of several processes, that is to the parallelism.

Consider the allocation of a resource of which there are two copies, and for which the processes are in competition. One copy of the resource can be allocated to a process only if it is not being used by another. This rule, which ensures the consistency of the resource and its use can easily be stated in a synchronization expression. Consider the state variable:

var nbfreeres : $0 .. 2$ *initialized to* 2;

The condition of use is therefore *nbfreeres* > 0, with the implicit assumption that this variable is accessed in mutual exclusion.

Another way of expressing the synchronization constraint is to introduce two variables that detect the state of the resource by evaluation of the differences between the numbers of allocations and releasing of its various copies:

var alloc, free : $0 .. + \infty$ *initialized to* 0;

the condition is therefore: *alloc* $-$ *free* < 2

95

There are several formulations for the expression of synchronization relative to a given rule, which are linked (here: *nbfreeres* $= 2 - (alloc\text{-}free)$).

In both cases the conditions for using a resource are distinguished from the constraint by passing from a strict inequality into an inequality in the broader sense. The condition *nbrfreeres* > 0 ensures that the constraint *nbfreeres* ≥ 0 always holds. A constraint defines a synchronization invariant.

In this chapter the case of synchronization constraints that can be expressed in the form of a system S of p linear inequalities is considered:

$$ S = \left\{ \sum_{i=1}^{m} \alpha_{ij}\, x_i \leq k_j \right\} \quad 1 \leq j \leq p $$

The coefficients α_{ij} and k_j are integer constants; the x_i are control variables, handled by the processes which define the synchronization constraints. The processes can handle them at will as long as the constraints defined by S are always satisfied, that is that the value of the invariant must be conserved. As a result, the actions of the processes leading to modifications of the x_i must be subject to conditions that block them whenever those conditions are not satisfied, and the modification of the x_i must be consistent. It is possible to represent the values that the x_i can take by the set of integer coordinate points situated in the convex polyhedron defined by S in $\mathbb{R}^m$.

Most practical problems of synchronization can be formulated with constraints of this type. Apart from the example above (*alloc* $- free \leq 2$), Chapter 1 introduced synchronization conditions that subject the producers and consumers to the constraint that there can be no more than n messages produced and not yet acted on.

In a centralized context the mutual exclusion of memory accesses constitutes the basic tool with which the synchronization problems can be solved. Once this technique has been mastered and has been provided in the form of such things as semaphores or monitors, synchronization problems present no major difficulties. The reader is recommended to consult works on centralized systems and parallel programming for further details on this subject.

In this chapter the elements of a methodology for distribution of synchronization conditions will be introduced. These elements are based on the type of the state variables x_i involved in the synchronization constraints.

Note: Certain synchronization constraints that are not expressed directly by linear inequalities can nonetheless be adapted by changing the appropriate variables. For example:

$$ x_1^2 + x_2 - x_3 \leq k $$

is equivalent to:

$$ xx_1 + x_2 - x_3 \leq k $$

after the change $xx_1 = x_1^2$. The new variables, here xx_1, must now be managed.

5.1.2. Types of state variable

The example of the message transfer protocol discussed in Chapter 1, and that of logical clocks in Chapter 2 can be seen as two specific synchronization problems. In both cases, the state variables x_i with which the global synchronization expressions are formed are integer variables, which monotonically increase through time. They count the 'start of execution (or termination) of a procedure p' events. The procedure p, (whether *produce* or *consume* (Chapter 1) or *date* (Chapter 2)), is associated with only one given site. The associated counters therefore show exact values at the corresponding site (and their possible copies show approximate values at the others). Distributing a global synchronization constraint of this type involves knowing how to distribute the expressions in which such variables are involved. This will be discussed in Section 5.2.

The solution of certain synchronization problems is based on integer state variables x_i, which, although each is associated with only one given site/process, have an arbitrary behaviour in their domain of definition. They can increase or decrease according to the events that take place. In Section 5.3 the distribution of the constraint expressed with such variables will be considered.

Finally, whatever their behaviour, there are state variables x_i that do not belong to any particular site: these are 'global variables' in the sense that their behaviour is governed by several events that occur at different sites. This is studied in Section 5.4.

5.1.3. Distribution of a constraint

It will be assumed in this chapter that there is one and only one process per site, which will allow the two terms to be treated as one. This simplification will lighten the explanation without reducing its generality.

Distributing a global synchronization constraint will consist of introducing:

(i) locally, auxiliary state variables (that is, not appearing in the initial global constraint);

(ii) globally, protocols (distributed control algorithms).

The variables and protocols must allow the processes to evaluate, in place of the condition defined by the global constraint to which they are subject, a local condition that represents them 'correctly'; that is, such that the two following conditions are satisfied:

—the global constraint will never be broken;

—possibility of deadlock in the system that did not exist previously are not introduced.

In other words, the distributed application of the global synchronization must guarantee that its semantics (safety and liveness) are retained.

Note: The replacement of a global condition (one that cannot be evaluated by a process) by a stronger local condition (one that the process can evaluate) is a frequently used process in control algorithms. Consider, for example, the branch and bound technique, often applied for optimization problems. The introduction of a heuristic function which, for each node in the tree of possible solutions, gives an under-estimate of the remaining cost, is based on the same idea: 'do not take a decision without having accurate information, as any other attitude may have unfortunate consequences.' This would be the case for a heuristic function which might sometimes be an over-estimate (and which might pass by the optimal solution). In the case of synchronization this would be a local condition, not always implying the global constraint.

5.2. Site-specific counters

5.2.1. Types of global constraint

5.2.1.1. THE PRODUCER–CONSUMER PROBLEM RECONSIDERED

In Chapter 1, counters were used to define relationships of global synchronization. We now take up the idea again; in the case of those associated with the procedures *produce* and *consume*, they give:

$$condition\,(produce) = \#init\,(produce) - \#term\,(consume) < n;$$
$$condition\,(consume) = \#init\,(consume) - \#term\,(produce) < 0.$$

Each of the counters relates to an event concerning *produce* or *consume* and each of these two procedures is associated with a given site: that of production and of consumption respectively. Each counter therefore has exact values at one of the sites: its 'site of origin'.

The site of production cannot evaluate the constraint associated with *produce* except in a purely local way. The solution given is based on the introduction of auxiliary variables: at a given site a particular variable will have an approximate value (because of the nearest transfer delay) of the counter of which it is the image, and which is located at its origin site. The introduction of two copies $\#cterm(produce)$ and $\#cterm(consume)$ allows the initial global constraints to be replaced by locally evaluable constraints, such that this evaluation is consistent with respect to the corresponding global constraint:

$$condition\,(produce) = \#init\,(produce) - \#cterm\,(consume) < n:$$
$$condition\,(consume) = \#init\,(consume) - \#cterm\,(produce) < 0$$

This example is simply a specific case of the results that will be generalized below.

5.2.1.2. A CANONICAL FORM

The global constraints of interest here have the following form:

$$condition\,(p) = \Sigma\,\alpha_i\,c_i < k$$

in which:

—only the case where $\alpha_i = \pm 1$ is considered.

—the c_i are monotonically increasing counters: each of them is associated with a given site, its origin site, and therefore counts events relative to that site.

—k is a constant.

The two global conditions associated with the *produce* and *consume* procedures are of this form, which will be referred to as 'canonical'.

A constraint of this type must be evaluated on the site P_i which uses the procedure. Among the counters involved in the constraint, those with P_i as their origin site have an exact value there. The others will be represented at P_i by their images, managed in such a way that the two properties stated in Section 5.13 will hold:

—no blocking;

—local condition true $=>$ global condition true.

Two cases arise according to the sign of the coefficient α_i which modifies the counter c_i in the global constraint. The case of a stable environment will be considered with no loss, duplication or desequencing of the messages.

5.2.2. Lower-bound values and delayed updating

To illustrate the method, consider two sites P_1 and P_2 with which are associated the counters c_1 and c_2 respectively. Any execution of the procedure p by site P_1 is subjected to the global constraint $c_1 - c_2 < k$. The values of c_1 and c_2 are exact at P_1 and P_2 respectively. An image mc_2 of c_2 is held at P_1 such that when the local constraint $c_1 - mc_2 < k$ is satisfied, the same is true of the global constraint $c_1 - c_2 < k$.

If it is ensured that at any moment $mc_2 \leq c_2$ the desired property will be achieved:

$$\left.\begin{array}{c} c_1 - mc_2 < k \\ mc_2 \leq c_2 \end{array}\right\} => c_1 - c_2 < k$$

The image mc_2 at P_1 must always be less than or equal to the exact value of counter c_2: the counter image must be a lower-bound value of it. But this occurs naturally if, when site P_2 increases c_2, it sends to P_1 an update message for the image mc_2, since the updating is slowed by the delay in transfer, the image mc_2 is always less than or equal to c_2 (equality occurring when there is no updating message in transit).

Generally speaking, at a given site P_i it is therefore possible to replace the counters equipped with a negative coefficient in the global constraint (and for which P_i is not the origin site) with delayed updated images. This does not introduce deadlock or inconsistency in the evaluation of local conditions.

5.2.3. Upper-bound values and anticipated updating

5.2.3.1. INCREASE OF THE UPPER-BOUND VALUES

Before going any further, it should be noted that the global constraints must introduce positive and negative coefficients associated with the monotonically increasing counters. If the coefficients were all negative, or all positive, the condition associated with the constraint would either always be true or always be false (starting from a particular moment, as a function of the value of k). This indicates that either there is no constraint or the constraint blocks the system: such possibilities are not considered.

Consider, in the context of the previous illustration, the site P_2 which must satisfy condition $c_1 - c_2 < k$ before executing a procedure q. Clearly P_2 cannot be satisfied with an image mc_1 of the counter c_1 with delayed updating ($mc_1 \leq c_1$), as before. The local condition obtained, $mc_1 - c_2 < k$, could be satisfied, although the global condition would not be.

Let Mc_1 be the image of c_1 introduced on P_2 to enable it to evaluate locally the condition $Mc_1 - c_2 < k$. For this evaluation to be correct, the following must be true:

$$Mc_1 - c_2 < k => c_1 - c_2 < k$$

which will be verified if Mc_1 is a majorant value of c_1 : $Mc_1 \geq c_1$. On a non-origin site, therefore, a counter modified by a positive coefficient must be replaced with an image which is an upper-bound value so that the locally obtained condition is correct.

The protocol that will ensure that at any moment $Mc_1 \geq c_2$ is, by analogy with the above, the anticipated updating. When P_1 wishes to increase the counter c_1 by 1, it must first broadcast to the sites that possess images Mc_1 a message that increments these; when it is informed that all the images have been increased by 1, it can proceed to the increase of c_1. (Several techniques are possible to inform P_1 that all the images Mc_1 have been increased: in the case of a ring, return to P_1 of the incrementing message, for example; or in the case of a complete graph, acknowledgements of the increase, etc.)

In a more general way, P_1 can broadcast an incrementing message for a ($a \geq 1$). Thus, when P_1 is informed of the increase of the images, $Mc_1 = c_1 + a$, which allows P_1 to carry out locally a increases of 1 without informing the sites with an image Mc_1. An incrementing message of a corresponds to the taking of a credits by P_1 in its counter c_1.

5.2.3.2. REDUCTION OF THE UPPER-BOUND VALUES

The protocol for taking credits ensures that the concrete image evaluated locally does not violate the global condition. Its effect on the liveness of the system remains to be considered. It may be suspected of leading to deadlock, but in this respect the previous case can be considered, with the condition associated locally with the procedure q defined and used on site P_2 being:

$$condition(q) = Mc_1 - c_2 < k$$

Let the counter c_2, initialized to 0, be increased during the execution of q, with the counters c_1 and c_2 such that there is an invariant $0 \le c_1 - c_2 \le k$. It follows that if the site P_1 takes k_1 credits with $k_1 > k$, then $Mc_1 - c_2 > k$, which will block the site P_2, since the condition can never be satisfied. The deadlock may be general if the conditions evaluated on the site P_1 involve images of counters with P_2 as their origin site, etc.

Several solutions are possible. The first consists of limiting the taking of credits to a maximal value. The calculation of this value may be complex if the concrete condition to which a procedure is subjected involves several images with upper-bound values Mc_i. This solution, however, presents the advantage of being static.

A dynamic solution consists of defining a reduction protocol for the maximum values. In the example, if the process P_1 has taken k_1 credits on counter c_1 and has used only $k_1' < k_1$, it can restore k_1'' credits with $k_1'' \le k_1 - k_1'$. The reduction protocol for the upper-bound values is therefore as follows: P_1 broadcasts a reduction message of k_1'' credits to the sites which have images Mc_1 of the counter c_1: (unlike the increase protocol, the decreases do not have to be acknowledged: P_1 knows the number of credits it has $k_1 - k_1''$).

This protocol allows system deadlock to be avoided.

5.2.4. An illustration

The global constraints generally involve more than two counters. The management of their images on the sites that involve them in the locally evaluated constraints takes place as explained:
—a modified image of a negative coefficient must be a lower-bound value: its updating is delayed;
—a modified image of a positive coefficient must be a upper-bound value: its updating is anticipated.

The example of the producer–consumer discussed in Chapter 1 and summarized in the introduction illustrates the first case: that of lower-bound values.

Example 1:
Consider, by way of illustration, an example that combines the previous two. Two counters c_1 and c_2, which have as their respective origin sites P_1 and P_2, are used to express a global synchronization constraint GC:

$$c_1 - c_2 < k$$

This single constraint subjects the executions of procedures p and q, usable respectively by P_1 and P_2:

$$condition(p) = c_1 - c_2 < k$$
$$condition(q) = c_1 - c_2 < k$$

(c_1 may, for example, be the counter $\#init(p)$ and c_2 the counter $\#term(q)$: the constraint $\#init(p) - \#term(q) < k$ is frequently encountered (cf. the producer–consumer): here the same global constraint applies to these two sites).

The distribution of this global constraint is simple, with the application of the stated principles. The following are found:

—at site P_1:

the counter c_1 and the minorant mc_2

the condition to evaluate $C1 : c_1 - mc_2 < k$

—at site P_2:

the counter c_2 and the majorant Mc_1

the condition to evaluate $C2 : Mc_1 - c_2 < k$

—the updating protocols ensuring that:

$mc_2 \le c_2$ (delayed updating)

$Mc_1 \ge c_1$ (anticipated updating).

The four possible cases can be examined according to the site concerned and the value of the condition:

—at site P_1 for procedure p:

- case 1 : C1 true

$$\left. \begin{array}{l} c_1 - mc_2 < k \\ mc_2 \le c_2 \end{array} \right\} => c_1 - c_2 \le k : CG \text{ true}$$

—case 2 : C1 false

$$\left. \begin{array}{l} c_1 - mc_2 \ge k \\ mc_2 \le c_2 \end{array} \right\} => CG : \text{true or false}$$

—at site P_2 for the procedure q:

—case 3 : C2 true

$$\left. \begin{array}{l} Mc_1 - c_2 < k \\ Mc_1 \ge c_1 \end{array} \right\} => c_1 - c_2 < k : CG \text{ true}$$

—case 4 : C2 false

$$\left. \begin{array}{l} Mc_1 - c_2 \ge k \\ Mc_1 \ge c_1 \end{array} \right\} => CG \text{ true or false}$$

Note that when the conditions are satisfied, the same applies for the global constraint. In the case where $c_1 = \#init(p)$ and $c_2 = \#term(q)$, it is not possible, when c_1 and c_2 are respectively false, that $c_1 - mc_2 = k$ and $Mc_1 - c_2 = k$. Note that the process P_1 is sequential and therefore there is as the most one execution of the procedure p at P_1 at a given moment. The same is true for q at P_2. In the contrary case (several instances of q in parallel) it would be possible to have $Mc_1 - c_2 \ge k$.

Example 2

Consider the following global synchronization condition associated with the procedure p used by the process P_1 (the variable c_i is a counter associated with site P_i):

$$\mathbf{condition}(p) = c_1 + c_2 - c_3 < k$$

Site P_1 will therefore manage x_1 and the two images with anticipated updating Mc_2 and with delayed updating mc_3. The local condition is therefore:

condition $(p) = c_1 + Mc_2 - mc_3 < k$

This local condition implies the global condition. The generalization of the condition to an inequality with any number of counters is clear.

In the case where the global synchronization constraint subjects the execution of p at P_1 to a condition consisting of several inequalities, the generalization is also easy. For example:

condition $(p) = c_1 + c_2 < c_3$ *and* $c_1 < c_4$

The site P_1 therefore manages c_1, Mc_2, mc_3 and mc_4 (if only c_1 has P_1 for its origin site): the condition is therefore the conjunction of two simple conditions.

5.2.5. Message loss

The implicit assumption has been made that the messages would not be lost. If this is not the case then, if the system can be deadlocked (with an image never being modified), the local conditions will not violate the global constraint false. Resilience to loss, duplication and desequencing of messages requires the introduction of the methods given in Section 1.4.: sequence numbers associated with the messages, and systematic retransmission when acknowledgement does not reach the transmitter of the message.

5.3. Variables with arbitrary behaviour

5.3.1. Assumptions and types of global constraints

Of interest now is the case where the global synchronization constraints, given in the form of a system of linear inequalities, involves variables x_i, each associated with an origin site, which are not necessarily monotonically changing counters—they can be increased or decreased (it is assumed here that this will be in steps of 1, without loss of generality).

In the section above devoted to the counter variables, a certain number of assumptions were made: the communication channels were reliable, so there was no loss, duplication or desequencing of messages. In this part, a solution will be given on the same assumptions, but these will soon be abandoned to allow for unreliable channels. The solution proposed will then be general (and will handle the particular case of counters in an unreliable environment).

5.3.2. Reliable environment case

When the environment is reliable a simple solution consists of considering the case of monotonically changing variables. Each variable x_i, with whatever behaviour, can be affected by two counters c_i and c'_i such that:

$$x_i = c_i - c'_i$$

An increase of x_i corresponds to $c_i := c_i + 1$;
a decrease of x_i corresponds to $c'_i := c'_i + 1$.

The conditions that ensure that the global synchronization constraints are satisfied are shown in terms of counters. Their actual evaluation is based on the technique of upper- and lower-bound values, given above.

The principle used is simple and powerful: it consists of applying changes of variable in order to return to known solutions (cf. integral calculus, where such changes of variable are very frequently used).

5.3.3. Principle of the general solution

The above principle is no longer applicable if the messages evolve in an unreliable environment; the solutions proposed with counters assume a reliable environment. To simplify the explanation, consider a case in which the constraints are expressed by inequalities with two variables and are of the form:

$$x_i \leq x_j$$

This involves equipping two sites P_i and P_j, the origins of x_i and x_j respectively, a cooperation protocol ensuring that the constraint $x_i \leq x_j$ is always satisfied. A generalization will be given for the case of more than two variables.

5.3.3.1. RESILIENCE TO AN UNRELIABLE ENVIRONMENT

Chapter 1 introduced tools and techniques for resilience to loss, duplication and desequencing of messages. When the two sites are to communicate the values of their respective variables x_i and x_j, the systematic association of sequence numbers with the messages carrying these values allows the reconstruction of the order of transmission on reception, as well as the elimination of duplicates. In addition, gaps in the numbering show that the corresponding messages have been either lost or overtaken by messages with higher numbers.

The systematic retransmission of messages (on the timeout of a guard delay, for example), joined with the sending of positive reception acknowledgements, enables the effects of loss of messages to be ruled out. These techniques will enter into the solution of the problem considered.

5.3.3.2. ENSURING COMPLIANCE WITH CONSTRAINTS

Once resilience to breakdowns is ensured, a protocol must be found to ensure the satisfaction of the invariant. Different types of actions on the variables must be distinguished according to their effect on this. Consider the variable x_i, handled by P_i. It is clear that decreases in x_i present no danger for the invariant $x_i \leq x_j$.

Increases of x_i, on the other hand, must be controlled so as not to invalidate it. The same applies for x_j: only its decrease requires a control—that is, involves global information.

In order to explain the control to be exercised, consider a system equipped with a global clock that is used to label the message transmitted. A procedure of this type has already been seen in Chapter 4, where the algorithm for calculation of a global state is derived from a centralized formulation. Remember that at any moment only a single event (transmission or reception of a message, or modification of a variable) is possible, and that the channels are reliable. These two assumptions will be relaxed to obtain the desired protocol: the first by introducing logical clocks that will order the events, the second by the techniques given in Section 5.5.3.1.

The operations to be controlled, incrementations of x_i by P_i and decreases of x_j by P_j, involve information that P_i alone or P_j alone does not have, so they are subjected to the reception of control messages. A message of this type transmitted by P_i (or P_j) at date t will be made up of two fields: the date of transmission t and the value of x_i (or x_j) at that date.

Consider the process P_i, which at date tr receives the control message (xxj, ts). If P_i knows that between ts and tr the process P_j has not received any control message (from P_i) it can conclude that at instant tr the current value of x_j is greater than or equal to the value received xxj (in fact, in the absence of any control message from P_i informing it of x_i, the process P_j can only increment the variable x_j). P_i can therefore increase x_i if the condition $x_i < xxj$ is satisfied. Such increases ensure that the invariant holds.

In a similar way, the decreases of x_j by P_j must be subjected to the arrival of a message (xxi, ts') at date tr': they are only possible if $xxi < x_j$ and if P_i has not received any control message between ts' and tr' (since x_i can only have been decreased).

Once these rules are established, it is important to provide an operational mechanism to allow each of the processes P_i and P_j to know, on reception at tr of a message transmitted at ts, that the other process has not received anything between ts and tr. This will be achieved if it can be ensured that the messages are transmitted and received one after the other. There will then be a maximum of one message in transit (and consequently the operations to be controlled will be carried out in mutual exclusion) (cf. Figure 1).

5.3.4. A protocol guaranteeing invariance

The principles of solution having been established, they must be made concrete in an implementation.

5.3.4.1. AN IMPLEMENTATION OF THE PRINCIPLES

The (abstract) global clock will be implemented by logical clocks. Their function is summarized as follows: clocks h_i and h_j, associated respectively with the

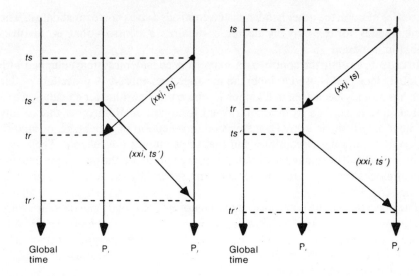

- *xxi* is not an upper bound of x_i at *tr'*
- *xxj* is a lower bound of x_j at *tr*

- *xxi* is an upper bound of x_i at *tr'*
- *xxj* is a lower bound of x_j at *tr*

Figure 1. Transfer of messages.

sites P_i and P_j and initialized to 0, are variables that may only increase: the management of their growth is adapted to the problem, as shown.

Each of the processes associates the value of its clock with any message it transmits: this is the logical date of transmission. Several messages can be transmitted with the same date: the receiver process must then consider only the first of these messages and disregard the others. (This means that if P_i, for example, transmits the messages (α, h_i) and then $(\alpha - 1, h_i)$ only the first to arrive will be taken into account by P_j: on receiving (β, h_i) and taking it into account, P_j knows that $x_i \leq \beta$.)

In order to allow for possible losses, the process P_i (resp. P_j) may at any moment (and must, from time to time) transmit the message (x, h_i) (resp. (x_j, h_j)).

As will be seen, the clocks, in addition to playing the role of implementing the global clock, also take on the role of the sequence numbers, consequently these do not need to appear explicitly. Resilience to instability of the channels is therefore ensured here by systematic retransmission and the use of clock values to label the messages.

It remains to be ensured that, in the logical time represented by h_i and h_j, at the most one message is in transit at a given moment (Figure 1). This can be achieved by comparing the date h of transmission of a message (x, h) with the value of the clock in the receiver. The comparison will depend on the receiver. If P_i and P_j carry out the same comparisons and modifications of their clocks, both will be able either to accept messages simultaneously or to refuse them (which will lead to deadlock of the system).

Consider the case of P_i when it receives a message (xxj, h) from P_j. P_i will not accept such a message unless $h \geq h_i$ (it rejects it in the contrary case). It then updates its clock by carrying out the synchronization $h_i := h+1$. (It will therefore reject any subsequent messages with dates less than or equal to h, and the messages it will transmit will have dates greater than h.)

Now consider P_j when it receives a message (xxi, k); it will not accept it unless $k > h_j$ and will therefore resynchronize its clock with $h_j := k$. The messages it previously transmitted have dates less than that of k and will be rejected by P_i. This ensures that between transmission at ts and reception at tr of the message (xxi, k) accepted by P_j, the process P_i has not been able to increase x_i. The same is true when P_i receives and accepts a message (xxj, h): the messages it has transmitted previously are rejected by P_j since they carry a date less than or equal to h. P_j has therefore not been able to decrease x_j since transmission of the message (xxj, h) (it has only been able to increase it).

Systematic transmission and retransmission, in conjunction with the symmetry of behaviour of the processes (rejecting and accepting messages) and the asymmetry in the management of clocks, ensure that at a given moment there will only be one control message handled by one of the processes (partial correctness) and there will effectively be one (so the system is not deadlocked).

Note: It would have been possible to introduce the asymmetry in a different way by, for example, introducing the site identities: for example, the couplet (h, i) associated with any message would allow the smallest number to be preferred, in conflicts. Such a solution, which is standard when the aim is to impose total ordering on the message (cf. Section 4.2.2), would be rather less cumbersome.

Figure 2 makes Figure 1 more concrete by illustrating the various conflicting situations that arise as a function of the clock values (note that these are not resynchronized except during the reception of messages).

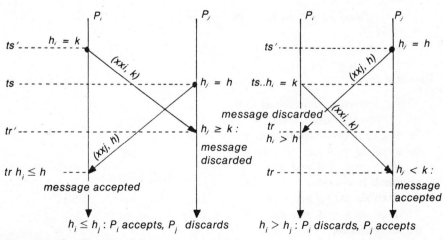

Figure 2. Conflict situations.

5.3.4.2. CONTROL ALGORITHMS

In order to give the two processes the greatest possible independence the reception of a message at P_i should be dissociated from the possibility of increasing x_i and at P_j from the possibility of decreasing x_j. For this, variables mx_i and mx_j are introduced into P_i and P_j respectively, indicating to them how far they can respectively increase x_i and decrease x_j without danger. There are therefore two local invariants: $x_i \leq mx_i$ and $mx_j \leq x_j$ and the global invariant $mx_i \leq mx_j$ which ensure that $x_i \leq x_j$ is respected. The two local invariants are guaranteed by the processes themselves and the global invariant $mx_i \leq mx_j$ is guaranteed by the previous protocol which operates on the variables mx_i and mx_j.

Protocol for P_i

var h_i : increasing initialized to 0;
  x_i : integer initialized to α;
  mx_i : integer initialized to β: ($\alpha \leq \beta \leq \gamma \leq \delta$)
on a call to decrease
  do $x_i := x_i - 1$ end do
on a call to increase
  possible only if $x_i < mx_i$;
  do $x_i := x_i + 1$ end do
on transmission
  always possible and performed regularly;
  do $mx_i := x_i$;
    send (mx_i, h_i) to P_j
  end do
on reception of (mxj, h) from P_j
  do
    if $h \geq h_i$ then
      $h_i := h + 1$;
        if $mxj > mx_i$ then $mx_i := mxj$ end if
      end if
    end do

Protocol for P_j

var h_j : increasing initialized to 0;
  x_j : integer initialized to δ;
  mx_j : integer initialized to γ;
on a call to increase
  do $x_j := x_j + 1$ end do
on a call to decrease
  possible only if $mx_j < x_j$;
  do $x_j := x_j - 1$ end do
on transmission
  always possible and performed regularly;

```
    do mxⱼ := xⱼ;
      send (mxⱼ, hⱼ) to P
    end do;
on reception of (mxi, k) from Pᵢ
    do
      if k > hⱼ then
        hⱼ := k;
          if mxi < mxⱼ then mxⱼ := mxi end if
      end if
    end do
```

A possible interpretation of mx_i (resp. mx_j) is to consider this variable as the approximate knowledge that the process P_i (resp. P_j) has of the value of x_j (resp. x_i).

As an exercise the reader is recommended to give the protocol in which reception of a value and the incrementation of x_j by P_i (and the decrease of x_j by P_j) are not dissociated: the variables mx_i and mx_j are of no use, and the control is expressed directly by x_i and x_j.

As a further exercise, to illustrate the operation of the algorithm, the reader may consider the limiting case in which $x_i = mx_i = \alpha$ and $x_j = mx_j = \alpha + 1$ with either $h_i \leq h_j$ or $h_i > h_j$ (in the first case P_i can increase x_i, while in the second it is P_j that can decrease x_j).

The event form given to the algorithm and the symmetrical management of variables mx_i and mx_j may introduce 'livelock' unless decrease and increase of x_i are made priorities (when possible) at P_i compared with transmission (always possible), and the increase and decrease of x_j are made priorities at P_j relative to transmission. If this were not the case, P_i and P_j could 'operate in a vacuum'; for example, in the case $x_i = \alpha = x_j - 1$, if the two processes each execute an infinite loop (reception–transmission). Another solution is to manage mx_i and mx_j in an asymmetrical way in the two control algorithms. Note that this problem does not arise if the increase/decrease is not dissociated from the reception of a message, since the variables mx_i and mx_j would therefore not exist.

5.3.5. Towards a solution of the general case

Consider the assumptions made in Section 5.3.3. In the general case, a global synchronization constraint may involve more than two variables linked to various sites. The general form of a constraint (in which the coefficients α_i are assumed to be equal to ± 1) is therefore:

$$\Sigma \, \alpha_i \, x_i \leq k$$

k is assumed to be equal to 0 without loss of generality (an additional variable initialized to k and never subsequently modified may simply be considered).

Let I and J be the sets of variables provided with coefficients respectively positive and negative. The previous constraint may then be expressed:

$$\sum_{i \in I} x_i \leq \sum_{j \in J} x_j$$

Since there is no known *ad hoc* solution to guarantee such an invariant in which the variables x_i can be increased and decreased, one method consists of considering problems to which solutions are known. If the case where I and J are reduced to a single element can be solved, one solution consists of adopting that situation. Among all the processes P_i such that $i \in I$, one chooses a particular process such as P'_i, with the same for J, such as P'_j. Each of these two processes determines the management of Σx_i or Σx_j respectively. The operations for increase of the x_i are carried out in mutual exclusion, managed by P_i: the decrease operations do not need to be controlled but P'_i must be informed (this information can be known from P'_i with some delay). The same applies for the x_j : P'_j manages their decreases in exclusion and is informed (with a possible delay) of their increase. The sites P'_i and P'_j therefore have at their disposal two variables sx_i and sx_j such that:

$$\Sigma x_i \leq sx_i \text{ and } \Sigma sx_j \leq x_j$$

Using the above protocol P'_i and P'_j can ensure that the invariant

$$sx_i \leq sx_j$$

is satisfied. The dangerous operations are carried out 'by procuration' by P'_i and P'_j on sx_i and sx_j respectively for the count of a P_i (increase of an x_i in exclusion) or a P_j (decrease of x_j in exclusion). Although lengthy, this solution deals with the problem.

Another solution consists of breaking down the initial inequality into several inequalities with two variables. For example:

$$x_1 + x_2 \leq x_3 + x_4 \tag{C1}$$

can be broken down into:

$$x_1 \leq x_3 + k \tag{C2}$$
$$x_2 \leq x_4 - k \tag{C3}$$

In fact, the two constraints introduced are such that:

$$C_2 \text{ and } C_3 \Longrightarrow C_1$$

These constraints with two variables cannot then violate the initial constraint. In order not to introduce any block, the value of the constant k must be chosen carefully. A protocol to redefine the value of k during the calculation may be envisaged between P_1, P_2, P_3 and P_4 (cf. the protocols for the management of the lower- and upper-bound values in Sections 5.2.2 and 5.2.3). The problem of the initial choice of k may be difficult as a single variable x_i is involved in several global constraints.

5.4. Non-site-specific variables

5.4.1. Principle of solution

In Sections 5.2 and 5.3 above we examined the protocols that the processes must obey when the global synchronization constraints, expressed in the form of linear inequalities, involve variables each of which is linked to a particular process. There are synchronization problems, however, the constraints of which involve 'global' variables—that is, variables not associated with any particular site. This is, for example, the case with the allocation of r resources, common to the whole system, and potentially needed by all processes. A formulation for the synchronization constraint for this problem is, for example: *nbfreeres* ≥ 0 with the variable *nbfreeres* initialized to r, and able to be increased and decreased by any process. A formulation in terms of counters, counting up the acquisition and freeing numbers (*alloc* and *free*) of the resources:

$$0 \leq alloc - free \leq r$$

does not change the problem at all: the two counters are 'global'.

The general principle for distributing such global synchronization constraints consists of introducing new variables to reduce the case to that of site-specific variables, possibly with an additional protocol to make the solution obtained more flexible. (This principle was used in Section 5.3.2., where the variables x_i were split between two counters c_i and c'_i. In Section 5.3.5. the dual principle was used: the variable sx_i is the 'gathering' of variables x_i.)

Note that the synchronization constraints in which each of the variables is linked to only one site can be formulated using variables without any origin site. $\Sigma \, \alpha_i x_i \leq k$ for example, can be expressed as $y_0 + y_1 \leq k$ by introducing two new global variables and making suitable changes. The aim is therefore, when such a global formulation is given, to take the opposite route by introducing variables linked to specific sites.

To illustrate this concept, the example concerned with the allocation of common resources, will be treated with the use of a variable of arbitrary behaviour, then with monotonically increasing counters.

5.4.2. Splitting up variables

Consider the case in which the system is made up of n sites, each of which may acquire and then free one of the r common resources, and consider the formulation of the synchronization constraint based on the counters:

$$condition\,(acquire) = alloc - free < r$$

The solution by splitting consists of splitting each of the variables *alloc* and *free* at the set of sites by introducing new counters *alloc$_i$* and *free$_i$*, counting at site P_i the allocation and freeing numbers of the resources effected at that site. The following relationships apply:

$$alloc = \sum_i alloc_i$$

$$free = \sum_i free_i$$

and the condition can be expressed with counters linked to specific sites:

$$condition(acquire) = \sum_i alloc_i - \sum_i free_i < r$$

The distribution of a global condition of this type is simple. Consider site P_i with application of the rules seen previously: the variables $free_j$ that have a negative coefficient are replaced by an image $mfree_j$ which is a minimum value ($mfree_j \leq free_j$) and those that have a positive coefficient $alloc_j$ are replaced by an image $Malloc_j$ which maximizes them ($Malloc_j \geq alloc_j$). The delayed and anticipated updating protocols respectively ensure the coherent management of the variables introduced. The concrete condition to which the acquisition by P_i of a resource is subjected is the following:

$$condition(acquire) = alloc_i + \sum_{j \neq i} Malloc_j - (free_i + \sum_{j \neq i} mfree_j) < r$$

(Note that the condition of freeing of a resource is always satisfied.)

This implementation introduces two variables linked to each site and $2(n-1)$ variables which are the images of variables linked to the other sites; this gives $2n$ variables per site. As there are n sites there are $2n^2$ variables introduced in total to represent two global variables.

The spatial complexity of this implementation of a constraint with two global counters is therefore $O(n^2)$. This can be onerous given the time taken by the updating protocols of the variables introduced.

5.4.3. Partitioning and isarithmic control

Another solution consists of starting with the formulation based on the unconstrained variable *nbfreeres*, and breaking it down into as many variables as there are sites, each one being linked to a single site. This is the same splitting process as before but here, as it is not a question of counters, the term partitioning is used.

The site P_i is therefore provided with the variable $nbfreeres_i$ initialized to r_i; with $\sum_i 4_i = r$. The concrete condition to which the acquisition of a resource by site P_i is subject, is therefore:

$$condition(acquire) = nbfreeres_i > 0$$

This solution consists simply of breaking down the initial global constraint into n independent local constraints. This type of static solution may, nonetheless, lead to the system being deadlocked: if r is less than n, since the resources are indivisible there are sites P_i in which the variable $nbfreeres_i$ will have 0 as its initial value.

A site of this type would therefore not be able to gain access to a resource. In order to overcome the disadvantages, a suitable protocol must be introduced. The processes must exchange resource use credits so as to avoid the blocking of some of them. This technique, known as isarithmic control in networks, allows the admission of messages to be controlled with prevention of any blockage. In the example considered, the sites may be connected in a ring and the following protocol introduced to avoid any blockage: at a given moment a site P_i may give the neighbouring site P_j all or part of its use credits (initially it has r_i); the latter then adjusts r_j and may undertake the same actions.

The complexity of this solution is of one variable per site. It is therefore $O(n)$ in total. The protocol for isarithmic management is simple. This solution is, accordingly, of more interest than the one mentioned previously. Another advantage lies in the bounded range of values of the variables, which is not the case with counters.

5.4.4. Duplication and exclusion

The two previous solutions (splitting up of counters and variable partitioning) make no assumption of mutual exclusion between various accesses to the state variables.

A third solution consists of duplicating each of the variables on each of the sites and:

—ensuring mutual exclusion between the accesses to various copies of the variables;

—guaranteeing that all the copies (of each of the variables) have the same value.

Once these two points are established, each site has a copy conforming to the global state, which allows it to change without any problems.

A solution of this type is onerous to apply and is essentially based on mutual exclusion in a distributed context. More information on this can be found in Section 6.4.4.

5.4.5. A specific case

One particularly interesting case, concerning the constraints of global synchronization, is discussed below. Consider the integer variable x (with arbitrary behaviour and no origin site), involved in the expression of a global invariant $I(x)$ of the system. For the sake of simplicity, a system made up of two sites P_1 and P_2 is considered, with access to x for reading and writing (the rules stated are easily extended to any number of sites).

The special features of the case considered are the following:

—as concerns the modifications of x : P_1 carries out only increases whereas P_2 carries out only decreases.

—the invariant I is of the form: I_1 *and* I_2 *and* K where I_1 (resp. I_2) involves only the local variables of P_1 (resp. P_2) and x, and K do not involve x.

—I_1 (resp. I_2) is a monotonic non-decreasing (resp. non-increasing) function of x.

In other words $I_1(x) => I_1(x')$ for all values of x' greater than x; and the same thing for I_2 with x'' less than x.

The aim is to implement the global variable x in a distributed way, so as to be able to use the invariant I to express the synchronization constraints. The variable x may be implemented by the two variables x_1 and x_2 situated respectively at P_1 and P_2 and initialized to the initial value of x. The following simple protocol will ensure that:

$$x_1 \geq x \geq x_2$$

is always satisfied and that when there are no control messages in transit between P_1 and P_2, then:

$$x_1 = x \text{ and } x = x_2$$

Protocol of P_1:

on incrementation of *(alpha)*
  do
    $x_1 := x_1 + alpha$;
    send *plus (alpha)* **to** P_2
  end do
on reception of *minus (beta)*
  do
    $x_1 := x_1 - beta$
  end do

Protocol of P_2:

on decrease of *(alpha)*
  do
    $x_2 := x_2 - alpha$;
    send *minus (alpha)*
  end do
on reception of *plus (beta)*
  do
    $x_2 := x_2 + beta$
  end do

It is easy to show that the predicate I', obtained from I by replacing x by x_1 in I_1 and x by x_2 in I_2, is an invariant (because of the monotonic nature of I_1 and I_2). The invariant I' and the relationship $x_1 \geq x \geq x_2$ can be used to express and distribute the synchronization constraints. Note that $I => I'$ but the reverse is not necessarily the case when the control messages are in transit.

It will be noted that because the operations on the variable x are commutative and associative, the control messages may be desequenced without causing any problems. There must, of course, be no loss or duplication.

5.5. References

The formulation of synchronization constraints in terms of linear equations has been proposed by several authors, in particular Muntean (1977) and Bochmann (1979). The expression of constraints using counters is introduced in Robert and Verjus (1977) and developed in André *et al.* (1983).

The problem of distribution of constraints in which each counter is associated with only one site is tackled by Bochmann (1979, 1983), who suggested lower-bound values with delayed updating. The generalization to the case of upper-bound values with anticipated updating is due to Herman (1981).

The distribution of synchronization expressions in the case of variables with arbitrary behaviour, and where messages may be lost, duplicated and desequenced, has been studied by Carvalho and Roucairol. They introduced the principle guaranteeing the invariant (1982), and applied it to the design of distributed algorithms (1983). The theory for this will be found in Carvalho (1985).

The techniques for splitting and partitioning global variables, based on the principles of decomposition, are studied and illustrated in Verjus (1983), André *et al.* (1983), and Plouzeau *et al.* (1987). Isarithmic control has been introduced into networks to avoid blocking due to message congestion (Price [1974]). These techniques and the associated protocols constitute bases for the distribution of global variables. The particular protocol in which a variable can be incremented only by certain processes and decremented only by certain others was developed by Chandy (1985).

The problems of synchronization in a centralized context are tackled in works devoted to centralized systems (Krakowiak [1985], Peterson and Silberschatz [1983]). Also worthy of note are Ben Ari (1982) and Schiper (1986), which concentrate on centralized synchronization, and Raynal (1984), which concerns mutual exclusion and its solutions, both centralized and distributed. The reader will also find in Alpern and Schneider (1985) a study of the properties of liveness in systems, allowing deadlock-free situations to be characterized, and in Lamport (1984a) a neat presentation of the problems of synchronization in a distributed context.

Chapter 6

Elements and algorithms for a toolbox

6.1. Introduction

6.1.1. The idea of the toolbox

The design and implementation of sequential applications can benefit from the use of many ideas and tools of a linguistic, methodological or technological (either software or hardware) nature. Two examples, on very different levels, are modular programming on the one hand and on the other, the use of guard values to prevent jumping out from a cycle before it has been completed. Another is the systematic association of data structures with the control structures to enable them to be used, for example arrays with iteration, lists or trees with recursion. It is obvious that an understanding of such elements is a prerequisite if suitable solutions are to be developed for the problems presented.

In the field of parallelism based on the sharing of a common memory, basic elements also appear both at process level and in the method used for managing their interactions. These elements are generally included in the definition of a language, which makes it easier to use.

The set of all such elements that have been validated both theoretically and experimentally constitutes a toolbox (with each tool being delivered with instructions) that must be understood and mastered if real solutions to problems are to be found. The same applies in a distributed context.

6.1.2. The elements to be studied

The aim of this chapter is twofold: firstly the introduction of techniques and tools adapted to the solution of problems in distributed applications and systems: Since the area of distributed systems is still in its development stage, this will not concern the provision of 'definitive' tools but rather the illustration, using a number of important elements, of some of the main themes in distributed systems design and implementation.

The second intention of this chapter is to analyse and study a number of distributed algorithms, using illustrations of the tools and associated techniques introduced, for the solution of specific problems encountered in applications and systems.

The chapter is divided into four parts. Section 6.2 is a summary of a certain number of algorithmic tools introduced and used in the previous chapters. Sections 6.3 and 6.4 concern two tools and the implicit properties associated with them. These tools can be expressed in terms of global knowledge given *a priori* on the communication network supporting the system or the application. This partly consists of temporal knowledge: the maximum time taken by the transfer of a message between two neighbouring processes. This information allows finite delays (a message that has not been lost progresses along a line at an unspecified positive speed) to be replaced by bounded delays (the minimal speed is known). The second element of global knowledge is topological (or 'spatial'): this concerns the structure formed by the network. A specific topology, a ring for example, allows specific controls to be applied. It will be seen that it is possible to associate specific controls with certain structures. These two elements of *a priori* knowledge constitute two software tools, the judicious use of which may facilitate the solution of problems. The last part (Section 6.5) concerns the use of another tool, also given in the form of knowledge held initially by the processes: the identities of their neighbours.

The examples given below will show that an explicit knowledge of the specifications on which the definition of a system is based is essential if all the properties attached to them are to be of benefit. Once the available tools and their characteristics are understood, through the specifications, finding a solution for a problem can be undertaken far more systematically.

6.2 Summary of tools introduced and used

6.2.1. Sequence numbers and acknowledgements

Sequence numbers were introduced in Chapter 1 to prevent duplication and desequencing of messages. The association of consecutive numbers with messages successively transmitted allows them to be identified with a view to their control.

The systematic retransmission of messages when guard delays are exceeded allows the loss of these messages to be avoided. With the use of reception acknowledgements carrying the sequence numbers of the messages acknowledged, this process offers a real solution to the problem of unreliability of the channels.

Section 1.5.2 explained how to limit the range of values of sequence numbers: if ct is the credit of the transmitting entity, the range can be defined by the interval $0 . . m$, as long as $m > 2ct$ and the delay in message transfer is less than the time taken for the assignment of a certain amount of distinct numbers.

These tools and the application assumptions that accompany them therefore allow reliable transfer to be carried out.

6.2.2. Counters and logical clocks

Counters record the number of occurrences of events in the systems with which they are associated. In Chapters 1 and 5 it was shown that they constitute a fundamental element on which the expression of synchronization constraints can be based. Only the use made of them distinguishes counters from sequence numbers: the former keep a count in order to achieve synchronization while the latter are used to identify distinct and consecutive messages.

Logical clocks, introduced in Chapter 4, are similar to the two tools mentioned above. Relative to a given process P_i, a logical clock associates distinct identifiers with each of the events of P_i to express the sequence of the events. The protocol for clock re-setting following the reception of a message allows this property to be extended to the events of several sites, with the identifier of a reception always indicating that it is subsequent to the corresponding transmission. In addition, if the identifier of an event is taken as the ordered pair formed by its date of occurrence and the identifier of the process which is its source, it is possible to order totally all the events in the system. If (h, i) and (k, j) are the identifiers for the events e_1 and e_2, the following relation *precede* may be defined:

$$(e_1 \; precede \; e_2) \; iff \; (h < k \; or \; (h = k \; and \; i < j))$$

(the processes all have, by definition, distinct identifiers).

A total ordering relation of this type is based on identities consisting of a static component (the location at which the event occurs) and a dynamic component (its *date*). Apart from progressing from a space–time view to an interleaving view (cf. Chapter 4), such relations are the basis of techniques for avoiding deadlocks and starvation in distributed systems. (These are properties that must be guaranteed if the system is to function correctly.) For this, the identity of the event created must be associated with the request for it; the requests are therefore totally ordered, and any ordering of the services that obeys the ordering of the identities ensures that the system will not deadlock.

From the point of view of methodology, logical clocks constitute an extremely powerful tool. Chapter 4 showed that they enable a distributed algorithm for the calculation of a global state to be derived from a formula based on a global clock. To achieve this, the properties of the global clock on which the initial formulation was based was explicitly developed, and constructed using logical clocks. The method is general, and easy to use.

6.2.3. Markers and tokens

The markers introduced in Chapter 4 are simple and powerful algorithmic control tools. When the messages do not overtake each other, for example, they allow the set of messages transported on a channel to be partitioned into two subsets: those transmitted before the marker and those transmitted after. In this use, the aim is to detect a temporal property of the messages: their occurrence relative to a particular event.

A tool of this type can be used in control algorithms. A single specific message in the system is often used to visit and mark the process: such a marker, generally called a control token, is frequently used.

6.3. Temporal aspect: bounded delay

6.3.1. Introduction

This section concerns the effect of *a priori* knowledge of an upper bound on the transfer time for messages on a communication channel. This knowledge constitutes a design tool for distributed algorithms. It will be illustrated with four distinct examples: finding a global state in the specific case of the detection of deadlock, mutual exclusion, termination, and distribution of a state machine. In all these illustrations, Δ will designate the maximum transfer time (assumed to be identical on all the lines). This is *a priori* global knowledge.

6.3.2. Avoiding false deadlock

6.3.2.1. DEADLOCK AND FALSE DEADLOCK

Consider a distributed system with resources (a single copy of each) situated at various sites. These resources must be accessed in exclusion; for this the sites/processes have at their disposal two primitives: *require* (*r*) and *free* (*r*) (*r* designates the resource in question).

The 'wait-for graph' is an oriented graph that evolves dynamically under the effect of the acquisition and freeing of resources. The vertices are the sites. Arc (P_i, P_j) is introduced into the graph when P_i asks for a resource *r* currently held by P_j; when P_j frees the resource *r* the arcs (P_k, P_j) are cancelled from the graph, if they exist. The resource *r* is then assigned to one of the waiting processes P_i and new arcs (P_k, P_i) are created as a consequence. The management of the graph and the operation of the resource allocator are not discussed further here.

Consider a centralized system in which the resource allocator manages the wait-for graph. Using the centralized context, the graph describes the state of the processes at any time from the point of view of the resources. A cycle in the graph therefore describes an effective blocking situation. Consider the cycle $(P_i, P_k, P_l \ldots P_j, P_i)$; P_i is waiting for a resource held by P_k, but the latter will not free it before having obtained a resource currently held by P_l, which is, itself, etc. The cycle indicates that P_i is dependent on itself, thus characterizing the deadlock.

In a distributed system (with a reliable system of communication) a solution to the problem of resource allocation and the detection of deadlock may consist of distributing the allocator over the sites where the resources are located, and placing the detection of deadlock on a particular site P_d. When a site P_i wishes to acquire a resource r, it asks site P_k, which manages it. If r is allocated to P_j, P_k informs P_d, which manages the wait-for graphs, and adds the arc (P_i, P_j). Site P_d therefore keeps a representation of the global state of the system relating to the detection of deadlock. Is the view of the global state held by P_d correct? In other words, does a cycle in the wait-for graph permit the conclusion that there is deadlock?

To answer this, consider the following configuration in which $(P_1, P_2 \ldots P_k)$ constitutes a directed path through the waiting graph managed by P_d. P_k transmits a message freeing the resource r_1 which it was using to the allocator site of r_1. This site $P(r_1)$ allocates r_1 to P_{k-1} and informs P_d of these modifications using the control message m_1. After having freed r_1, P_k requests a resource r_2 from site $P(r_2)$ which manages it, but this resource is held by P_1: site $P(r_2)$ therefore signals to P_d, using the control message m_2, the (blocking) request from P_k for r_1 held by P_1 (Figure 1).

If the message m_1 arrives at P_d before m_2, the detector site will treat them in that order and will carry out the modifications they indicate on the representation it has of the global state, that is: $(P_1, P_2 \ldots P_{k-1})$ after the reception of m_1 and $(P_k, P_1 \ldots P_{k-1})$ after that of m_2. If, on the other hand, the messages arrive in reverse order (and there is nothing to prevent this happening, since the assumption con-

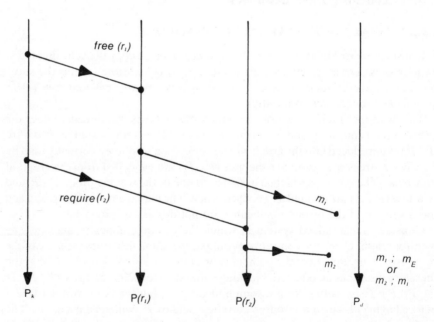

Fig. 1. Messages in transit.

cerning desequencing bears only on sites taken in pairs) the modification brought about by m_2 produces the cycle $(P_1, P_2 \ldots P_k, P_1)$ and P_d concludes that there is a deadlock that does not exist. This situation is called 'false deadlock'. A cycle in the wait-for graph is therefore not a sufficient condition to indicate deadlock. This is due to the approximate perception that P_d has of the real global state. Waiting for the arrival of m_1 before deciding is of no use: P_d does not know that the message is in transmit (and if m_1 does not exist, the deadlock is real).

6.3.2.2. A CONSISTENT GLOBAL STATE

One solution of the problem consists of using the global state calculation algorithm introduced in Chapter 4. Since deadlock is a stable property for the systems and the calculated global state is consistent, the detector process P_d may initiate global state calculations, and each time it obtains the local states of the processes and the states of the channels forming a global state, it can construct the wait-for graph from the global state and reach the conclusion that there is deadlock if the graph forms a cycle.

The stated solution, although correct, is based on a very general algorithm and consequently may be lengthy to apply for the specific problem of deadlock. The problem consists of offering process P_d a global state which may be only an approximate image of the real global state, but which will not be inconsistent.

Consider the assumption of a maximum delay Δ in the transfer of a message between two sites connected by a channel. To simplify matters, it is assumed that the actions carried out by the processes are of duration zero (non-zero durations can be taken into account by inclusion in Δ). A single time reference will be used, which has no physical existence but clarifies the operation of the algorithm, without participating in it.

Consider the principle of the algorithm introduced in Section 6.3.2.1, in which a process P_d maintains a representation (possibly approximate) of the global state from the point of view of deadlock—that is, a wait-for graph. Each site manages the resources located in it. When a process/site P_i wishes, at instant t, to use (resp. free) a resource r, it sends a message *require*(r) (resp. *free* (r)) to the site that manages r; that site receives the message at the latest at $t + \Delta$; once it has received such a message, it informs the detector site P_d. P_d is therefore informed at the latest at $t + 2\Delta$ of the requests and freeing that take place at t. It is assumed, to simplify matters, that P_d is directly connected to all the resource allocator sites.

Consider an edge of the wait-for graph managed by P_d (this graph gives an approximate global state of the blocking relationships) which has existed continuously for more than 2Δ: if P_d notes this at instant t it can conclude that the edge existed in the real wait-for graph at $t - 2\Delta$; (in fact, if it exists, the destruction of such an edge, not yet perceived by P_d at t, can take place only after $t - 2\Delta$). In general, if P_d notes that at instant t there is a cycle formed of edges which have all existed continuously and concurrently for more than 2Δ, it can conclude that this cycle existed at $t - 2\Delta$ in the real wait-for graph and, therefore, that there is deadlock. The interval 2Δ defines what is known as an observation period.

The sufficient condition for detecting a deadlock therefore consists of two component conditions: the presence of a cycle in the wait-for graph and the continuous existence of this cycle for at least 2Δ. The second of these is based on two assumptions. One is linked to the support: prior knowledge of the delay Δ by the detector processes P_d; the other is linked to the problem of deadlock itself: deadlock is a stable property.

The use made of the assumption Δ to obtain a global state pertinent to deadlock may be extended to the case of obtaining specific global states in other problems. The solution will also be dependent on the properties of the intended goal.

6.3.3. Mutual exclusion and order of satisfaction

6.3.3.1. A MUTUAL EXCLUSION ALGORITHM

Consider a set of n processes linked by a completely connected, reliable network. Each of the processes is offered a protocol allowing it to enter and leave a critical section—that is, one which is accessible to only one process at a time. The protocol must be free from deadlock and starvation, and should guarantee an order of satisfaction (cf. Section 6.3.2.2) between the requests for entry into the critical section by the candidate processes. For the time being it is assumed that delays are finite and undetermined.

Since mutual exclusion is a problem of assigning a unique privilege, one solution consists of indicating the privilege by means of a single message, known as a token (cf. Section 6.2.3). Since the algorithm must be fair (and fifo), the token will be made to circulate in a virtual ring formed by the processes requesting the critical section. A process requests the section by broadcasting a request to all the others (the graph being complete). A process receiving a request knows that the others have received it, or will do so (this would not be the case in an arbitrary network). To identify its successive requests, a given process will associate sequence numbers with them. The single token, in the possession only of the process that has the privilege, may store information relevant to the part of the global state effective in the allocation of the critical section. It will therefore store, for each process P_i, the number associated with the last request from P_i that has been satisfied. The updating of this information will be carried out by P_i itself in mutual exclusion when it has the token and before transmitting it. In order to establish the logical ring of requesting processes, each process P_i will maintain, relative to all the others, the numbers of their last requests. When P_i must transmit the token, it will transfer it to the first in the series $P_{i+1}, P_{i+2}, \ldots P_n, P_1, \ldots P_{i-1}$ such that the number of its last request received is greater than the number of its last request satisfied (stored in the token). This depends on the perception that P_i has of the state of each of the others relative to the mutual exclusion.

Each process is given the following declarations:

var *numseq*$_i$: 0 .. + ∞ *initialized to* 0;
    tokenpresent$_i$: **boolean**;
    inside$_i$: **boolean initialized to** *false*;
    copytoken$_i$: **table** [1 .. *n*] *of* 0 .. + ∞ *initialized to* 0;
    lastreq$_i$: **table** [1 .. *n*] *of* 0 .. + ∞ *initialized to* 0

The boolean *tokenpresent*$_i$ is initialized to false in all the processes but one; *inside*$_i$ is at true when P_i is in a critical section; *copytoken*$_i$ stores the value of the token when it is in P_i; *lastreq*$_i$ stores the perception that P_i has of the last requests transmitted by the others: thus *lastreq*$_i$ [*j*] contains the number of the last request from P_j perceived by P_i.

There are two types of message, *token* and *request*. Associated with the former is the value of the token; with the second, the identity of the transmitter and the corresponding sequence number.

The protocol for each of the processes P_i is as follows:

on *entry into a critical section*
  do
     if ⌐ *tokenpresent*$_i$ **then**
      numseq$_i$:= *numseq*$_i$ + *1*;
      broadcast request (*numseq*$_i$, *i*);
      wait for token (*copytoken*$_i$); —*interruptible wait*
      tokenpresent$_i$:= *true*
     end if;
     inside$_i$:= *true*
  end do
on *exit from critical section*
  do
     copytoken$_i$ [*i*] := *numseq*$_i$;
     inside$_i$:= *false*;
     for *y* **from** *i*+1 **to** *n*, 1 **to** *i*−1
       do
         if *lastreq*$_i$ [*j*] > *copytoken*$_i$ [*j*] **and** *tokenpresent*$_i$
          then *tokenpresent*$_i$:= *false*;
             send token (*copytoken*$_i$) **to** P_j
         end if
       end do
    end do
on *reception of request* (*k*, *j*)
  do
     if *tokenpresent*$_i$ **end** ⌐ *inside*$_i$
      then < *analogous text at exit from critical section*>
     end if
  end do

It is easy to show that this algorithm ensures exclusion (since the number of variables *tokenpresent*$_i$ with value *true* is at the most one), and is free from deadlock. It is also fair—that is, free from starvation—if the messages are delivered in a finite time. The execution by P_i of its protocol for exit from the critical section transmits the token to the first process P_k (with $k = i+1, \ldots i-1$) from which it has received an unsatisfied request. Since the messages are delivered in a finite time the processes are informed of the requests in a finite time and the requests will be honoured at the end of a finite period (on the assumption that any process that enters the critical section will remain there only for a finite period).

The algorithm requires 0 or n messages to achieve mutual exclusion.

6.3.3.2. ORDER OF SATISFACTION OF REQUESTS

In relation to an abstract global time, the requests have a certain total ordering. This ordering cannot be detected by any of the processes because of the distributed nature of the system. The total order of satisfaction of the requests is unconnected with the order in which they are produced (with the exception of the requests issuing from a single site which will be in the same order in both cases, since a process cannot make a new request until the previous one has been satisfied). There may be $n+2$ different orders: the order in which the requests are produced, the n possibly different orders in which the n processes perceive them, and the order in which they are satisfied.

Ensuring that the orders of perception and satisfaction are identical to the order of request production demands a global physical clock and therefore a centralized component to the algorithm.

One question that may arise is the following: what is the contribution of *a priori* knowledge of a maximum delay Δ, and how can this be used? Any process with such knowledge knows at instant t all the requests transmitted at or before $t - \Delta$, that is, the state of the system at $t - \Delta$ (certain of the requests transmitted were satisfied between t and $t - \Delta$). When P_i exits from the critical section at t, it will pass the privilege to the first of its neighbours on the logical ring of requesting processes $P_{i+1}, P_{i+2}, \ldots$ which integrates all the unsatisfied demands made at or before $t - \Delta$ and those made between $t - \Delta$ and t, and which have already arrived.

Such an assumption is therefore necessary if the maximum waiting time is to be quantified; if, in addition, the maximum time during which any process occupies the critical section is known, it is possible to calculate this waiting time explicitly.

Before concluding the discussion of this algorithm, the set of tools on which it is based should be mentioned: sequence numbers; *a priori* knowledge of the complete graph; construction of a logical ring; representation of a property by a special message, the token; representation of the global state pertinent to the problem viewpoint, partly in the token (as concerns the past relevant to the future) and partly in the process (as concerns unsatisfied requests).

6.3.4. A termination computation

6.3.4.1. THE PROBLEM

The problem of termination is essential to distributed control. Given a distributed algorithm, which will be called the underlying computation, the aim is to graft on to it an algorithm that will indicate its termination. The simpler the problem in a centralized context, the more complex it is when distributed.

This problem was tackled first in Section 4.4.1, where it was seen that since the termination property was stable, a solution to the problem could consist of repeatedly testing the property on a calculated global state until it is verified.

Consider the following: initially all the processes of the underlying computation are passive, with the exception of one of them, called P_0. A process may not become active until it receives a message, but may then send messages to other processes and possibly activate them. A process may transmit only a finite number of messages. In addition, to simplify matters it is assumed that P_0 can only transmit messages, and only at the beginning of the computation. This type of communication and computation is called 'diffusing computation' and it allows a large number of concrete problems to be modelled in which P_0 plays the role of the environment that starts the activity of a computation in the system.

6.3.4.2. PRINCIPLE OF THE SOLUTION

The tools on which the above solution is based have a particular control topology (the tree), and the communication delays are bounded by Δ.

When a process is activated by the reception of a message, it puts itself, if it has no parent, in a control tree with the message transmitter as parent. It informs the transmitter of this parental relationship, and the transmitter includes it among its sons. When a process is passive and has no sons it leaves the tree by informing its parent of the destruction of the parental relationship. Thus the active processes or parents of active processes form a tree with root P_0 (a process exists at most once in the tree). The control tree may develop and change its configuration during the calculation as a function of the active or passive state in which a process is at a given moment, and of the messages exchanged.

The principle of detection of termination is the following: when the process P_0 no longer has sons, there is no more activity in the underlying calculation, which is therefore terminated.

One problem remains to be solved: that of the perception that a process has of the number of its sons. In a centralized context there is no problem; the perception of the tree and its modifications are done in mutual exclusion. In a distributed system the transfer of messages takes time. Consider a process P_i which has no son in the tree and sends a message that will activate P_j and introduce it into the tree as a son of P_i. After sending this message P_i will become passive. The process P_i must therefore not conclude too hastily that it has no son in the control tree: it should wait for a possible control message from P_j to reach it, informing it of a parental relationship. If the processing time is considered to be zero (it can be 'absorbed' into Δ), P_i must wait 2Δ (a return journey for the message) before reaching its conclusion. It will then know whether P_j is one of its sons or not.

6.3.4.3. THE ALGORITHM

The aim is the control of a process P_i in the underlying computation. This control must not disturb the computation, and involves providing P_i with the following declarations (the processes have the identities 1 to n):

```
var state_i : (active, passive);
     nbsons_i : 0, . . n−1 initialized to 0;
     parent_i : 1 . . n ∪ {nil} initialized to nil;
     timeout_i : boolean initialized to true;
```

The behaviour of the control associated with P_i is as follows: the *control*-type messages have the value *exit* or *enter* (in the control tree).

```
on waiting or end
   do state_i := passive end do
on reception of message (m, j)
   do if parent_i = nil then parent_i := j;
                         send control(enter) to P_j
      end if;
      state_i := active
   end do
on desiring to transmit message(m, i) to j
   – – we have the state_i active and parent_i ≠ nil
   do
      send message(m, i) to P_j;
      armclock (2 Δ);
      timeout := false
   end do
on clock signal
   do timeout := true end do
on reception of control(m)
   do
      case m = enter → nbsons := nbsons + 1
           m = exit  → nbsons := nbsons − 1
      case
   end do
on transmission of control(exit)
   possible only if state_i = passive and nbsons_i = 0 and timeout;
   do
```

$x := parent_i;$
send $control(exit)$ **to** $P_i;$
$parent_i := nil$
end do

6.3.4.4. CHARACTERISTICS OF THE ALGORITHM

The algorithm is important because the values of the variables it involves are bounded, notably *nbsons*, which varies between 0 and n; unlike other algorithms, this one does not involve counters that accumulate the number of messages and the number of acknowledgements received by a process. The size of the variables is controlled statically (that is, at the compilation stage).

Waiting for a delay of 2Δ allows a process to reach a firm conclusion on the number of its sons: the perception it has when it can tests *nbsons* and finds it to be 0 is therefore exact (the boolean *timeout* prevents any test of *nbsons* when it is *false*).

This algorithm illustrates the use of Δ. A similar algorithm without the use of a time limit of this type may be envisaged. For this the principle of tree formation, explained in the algorithm for parallel traversal through a network (Section 3.3.2.2), is used. When P_i sends a message to P_j it includes the latter *a priori* on its list of sons. On reception of the message, P_j, if it is already in the control tree, replies that it should be taken out of the list (it will already have a parent). This principle is the dual of that in the algorithm presented. The inclusion of a process P_j in the control tree, as perceived by P_i, the transmitter of a message, may therefore be made by P_i:

—*a posteriori* after a delay of 2Δ after a transmission. If at the end of this delay P_j has not replied to P_i, the latter will not consider P_j to be one of its sons. It may therefore have fewer control messages exchanged than data messages, at worst twice as many (one *enter* and one *exit* message for each exchanged data message).

—*a priori* at transmission, P_j may reply to P_i to be removed from the set of P_i's sons, if P_j is already in the tree (cf. Section 3.3.3.2). There will therefore be only *exit* control messages, the same in number as the data messages exchanged by the underlying calculation. In addition, the variable $nbsons_i$ does not count the number of potential sons of P_i but the number of messages transmitted by P_i and not yet 'acknowledged' by *exit*. (This number is not bounded *a priori* since the process P_i can send several messages to a single process P_j without knowing whether P_j is one of its sons at the instant considered.)

The reader is recommended to give the corresponding algorithm (for the *a priori* case) and to study both algorithms in the following case. At t P_i sends a message to P_j and sends it another at t' with $t' > t$. At the first reception, P_j is in the control tree, but is no longer there at the second. It is interesting to examine i) in the first algorithm the two cases $t' < t + 2\Delta$ and $t' > t + \Delta$, and ii) in both algorithms, the phenomenon of desequencing.

The aim is to obtain an algorithm in which the range of variation of the variables is bounded and the transfer times are not. Each of the two algorithms above respects one of these two constraints. The problem of finding an algorithm that respects them both remains open.

6.3.5. Distributing a finite-state machine

6.3.5.1. THE PROBLEM

Many control problems can be resolved by expressing their solutions as finite-state automata (or machines). When an event occurs the automaton modifies its tables (state change) and may produce new events. It is therefore easy to express synchronization controls using automata.

Independently of the centralized or distributed character of the system, the control to be carried out is expressed in terms of a global abstract automaton A (cf. the expressions of global synchronization that are formulated independently of the centralized or distributed context). Once A is constructed, it must be distributed over the various sites of the system. For this, a copy a_i is produced at each of the sites. For these copies a_i to represent A correctly, each of them must have the same perception of the system—that is, perceive the same events in the same order. If the copies are activated in the same initial states and perceive their environment in the same way, they will evolve identically and produce the same results.

The problem therefore consists of coordinating the machines a_i so that all will receive the same inputs and treat them in the same order.

6.3.5.2. PRINCIPLES OF THE ALGORITHM

A preliminary solution consists of constructing a total ordering on each of the inputs (requests, for example), with each automaton a_i using these in the single order thus defined. It was seen in Section 4.2.2. that logical clocks allow this type of global order to be established when the following pair is associated with any request, identifying it without ambiguity: value of the clock of the transmitter process and identity of this process. (Cf. Section 6.4.4. below.)

Another solution consists of synchronizing the local automata a_i, using the knowledge that there is a maximum delay Δ if such is the case. At instant t, any site P_i broadcasts a message $m_{i,t}$ to all the others. At instant $t + \Delta$ each a_i will have received a message $m_{1,t} \ldots m_{n,t}$ from each of the sites: it can then consume the inputs in that order. Each copy a_i therefore perceives the same inputs in the same order: at $t + \Delta$ there is a global state (from the viewpoint that concerns it) of the system relating to instant t.

The principle of the algorithm consists of constructing a machine of the MISD type (Multiple Instruction Single Data), with each processor consuming the inputs and the unit behaving in a highly synchronous way.

6.4. Structural aspect: specific topologies

6.4.1. Introduction

The effect of topology on algorithms will be studied in this section, in terms of its effect on their design and properties. It has already been seen that certain solutions to problems are based on a particular graph; the algorithm of mutual exclusion and the distribution of a finite-state machine, seen above, are both based on complete graphs. What would happen if the graph were ring-shaped?

Three particular structures will be studied: the ring, the tree and the complete graph. Other structures can, of course, be envisaged (the hypercube of degree n, for example, in which any site has n neighbours and for which the diameter is n).

6.4.2. The ring

6.4.2.1. GENERAL REMARKS

The importance of the ring lies in its simplicity. Any process (or site) in it is connected to two other processes (the neighbours on the right and left), with the global structure including all the processes. Each process is simply the neighbour on the right or left of one of the other processes. A ring may be uni- or bidirectional depending on whether the messages can move in one direction or two.

A structure of this type can easily be used with special messages of the token type. An initial property of the structure is the following: when a message transmitted by a process P_i comes back to it (in a unidirectional ring), it will have visited all the processes, however many there are. It is therefore an obvious choice to construct learning algorithms on a ring shape, whether for learning the number of processes or their identities. A visit of this type to the processes is fair (it does not forget any process).

A priori knowledge of global information concerning the communication structure makes it easy to obtain other information. It is possible, for example, to construct an algorithm for the computation of a global state on a unidirectional ring. At a given moment, one or more processes initiate markers on the ring, record their respective local states, and start to record the messages received. On reception of one of these markers, a process that has not initiated any passes it to the next one on the ring, and records its local state and that of its input channel as empty. When a process that has initiated a marker receives one, it considers the state of its input channel to be the messages it has stored up to that point. If a single process P_i initiates a state computation, the marker can collect the local states of the processes visited, with all the channels being considered empty in the corresponding global state with the exception of the input channel of P_i, for which the process stores the messages received between transmission and reception of the marker.

The same applies in election algorithms for unidirectional rings. Initiating an election consists of a process transmitting its identifier to its single neighbour. On reception of such a message, a process passes it on to its own neighbour if its own identifier is smaller; if it is not, it initiates an election, if it has not already done so. Given the structure, it is easy to see that the process with the largest identifier will eventually initiate an election as soon as an election takes place, and that it will be the only message to go all round the ring.

Three problems are now considered in detail, with the effect of a ring structure on their solutions.

6.4.2.2. MUTUAL EXCLUSION

A simple mutual exclusion algorithm in a reliable ring consists of circulating a single token indicating privilege. Since the token always moves in the same direction, fairness is ensured. When a process receives the token it passes it on to the next process, if it does not wish to enter into mutual exclusion; in the contrary case, it enters into the exclusion and retains the token until leaving.

To make the ring structure and problems of exclusion clearer, a solution will be presented in which the processes have state variables and any process P_i can read the state variables of its left-hand neighbour P_{i-1} ($i-1$ *mod* n; the processes are arranged in order of their identities in the ring: $P_0, P_1, \ldots P_{n-1}, P_0$). It follows that the reading is achieved using messages transporting requests from P_i to P_{i-1} and the values of P_{i-1} to P_i. The ring is therefore bidirectional.

A predicate for exclusion

Let each of the processes P_i be provided with a state variable s_i. Then P_i has access to s_i for reading and writing and to s_{i-1} for reading (the operations on the indices representing the site identities are carried out *modulo* n). Therefore each process P_i does not know the global state but only part of that state, s_i and s_{i-1} (the knowledge of this latter variable may be delayed). The basic idea of the solution consists of finding a relationship PE between s_i and s_{i-1} such that it is satisfied only for a single pair (s_i, s_{i-1})

$$\exists ! \, i : PE \, (S_i, S_{i-1}) \text{ and } (\forall j \neq i : \neg PE \, (S_j, S_{j-1}))$$

PE represents the property of exclusion: at any moment the single pair of state variables that satisfies it defines the process P_i which has the exclusion.

A simple implementation of this idea consists, for example, of taking for PE:

$$PE \, (s_i, s_{i-1}) \equiv (s_i - s_{i-1} = k)$$

by considering that the s_i have integer values. A predicate NPE can therefore be found such that $NPE => \neg PE$; $NPE \, (s_j, s_{j-1})$ indicating to P_j that it is not preferred. (The same predicate NPE is chosen for all the sites which are not preferred in the interest of simplicity and regularity, to produce reproducible behaviour.) Consider the predicate NPE:

$$NPE \, (s_j, s_{j-1}) \equiv (s_j - s_{j-1} = k')$$

taking $k \neq k'$, this gives

$$NPE\ (s,\ s_{j-1}) => \neg\ PE\ (s,\ s_{j-1})$$

In other words, the state variable of each of the processes presents a shift of k' relative to the previous one, except for a particular process where the shift is k.

From the rules to the algorithm

Since the global exclusion property has been represented in the state variables, the algorithm must ensure that when P_i abandons the privilege (which it cannot do except by modifying s_i), it transmits it to P_{i+1}. The privilege will therefore move around the ring in the direction P_i, P_{i+1}, P_{i+2}, ... The property must also be true initially.

Consider the following 'simple' initialization: $s_i := i$: it ensures that only P_0 has the privilege by taking $k' = 1$ and $k = 1 - n$. Note that the identities of the n processes are $0, 1, \ldots n-1$. P_0 must therefore abandon the privilege by setting s_0 at a value such that the two following equations are satisfied:

$$s_0 - s_{n-1} = 1 \ and \ s_1 - s_0 = 1 - n$$

(Note that any process can modify only its own state variable.) This will be achieved with the modification $s_0 := s_0 + n$. The privilege is therefore held by P_1 and so on.

The algorithm obtained is the same for all the processes P_i (only the initializations of the s_i differ):

> *wait for* $(s_i - s_{i-1} = 1 - n)$;
> *<critical section>*
> $s_i := s_i + n$;

Other choices for the predicate PE (and the expression of $\neg PE$) are possible and lead to other algorithms.

Bounded variables

One drawback of the above algorithm is the infinite increase in the state variables s_i. The problem is how to bound the range of variation of the s_i.

Let the test and modification that concern s_i be carried out ***modulo*** p; what property must p satisfy for these operations to be possible? To answer this question consider the principle on which the algorithm is constructed: k must be different from k', so:

$$(k \neq k') \ \textbf{\textit{mod}} \ p$$

that is, $(n \neq 0) \ \textbf{\textit{mod}} \ p$. In other words, p must not be a divisor of n.

The final algorithm for P_i is therefore:

> initialization: $s_i := i \ \textbf{\textit{mod}} \ p$;

The entry and exit protocols for the critical section are:

waitfor $(s_i - s_{i-1} = 1 - n)$ *mod* p;
$<$*critical section*$>$
$s_i := (s_i + n)$ *mod* p;

This algorithm is symmetrical in the sense that the texts of all the processes are identical. The initializations are, however, asymmetrical.

Loss of messages

Compared to the algorithm based on a token, the present algorithm is of interest in the case where the ring may lose messages: the loss of the token representing the privilege does not endanger the exclusion, but requires the addition of an algorithm to detect the loss and, if necessary, regenerate the token. The algorithm considered here does not need this type of control: the privilege is in fact represented by the state variables. The implementation of the primitive *waitfor* repeats requests to read s_{i-1} until the test is positive; the loss of request messages can be compensated for by having P_i make them regularly while the value of s_{i-1} prevents it from progressing.

The algorithm given is based on a ring topology (in which the processes are arranged in order) to define the global state perceived by a process P_i. From this has been derived the predicate PE characterizing the exclusion and an application has been 'derived' from it. This principle can be used for other problems such as the allocation of resources in a ring.

6.4.2.3. TERMINATION

The problem

From an operational point of view, the problem of termination can be posed in the following terms:
  i) visit all the processes (and observe them to be in a passive state);
  ii) visit all the channels (and observe them to be empty);
  iii) carry out all these visits and assemble all the corresponding observations so that the global observation obtained (therefore defining a global state) is consistent.

If the global observation obtained is not consistent, it might be concluded that the underlying calculation has been terminated when this is not the case. This is called a false termination, and a correct algorithm should not indicate it.

The processes are connected by a network, unspecified *a priori*, defined by the underlying calculation. The solution of i) requires the introduction of a variable *state*$_i$ which indicates whether the process P_i is active or passive. The visits to processes can be carried out in various different ways: one of the simplest consists of going to each process in succession. The natural way of achieving this is to have an additional network, a uni-directional ring that includes all the processes; a special control message or token can then carry out an ordered visit of all the processes.

Operation ii) can be achieved with a simplifying assumption: the processes communicate in the underlying calculation or in its control by a rendezvous (during an exchange, the first of the two processes—transmitter or receiver—to arrive at the communication point waits for the other). The channels are therefore always empty. This simplifying assumption may be considered as unrealistic, and will in fact be abandoned later.

Rules of behaviour

To tackle operation iii), consider movement of the token through the control ring. Launched by a process, P_0 for example, in the passive state, the token behaves in the following way: a process that receives it does not transmit it to the next one until it is in the passive state. When the token returns to P_0 it has therefore visited all the processes (because of the ring structure) and observed them all to be in the passive state (because of the rule for circulation of the token). Can P_0 conclude that termination has occurred—in other words, does the composition of the local state observed by the token give a consistent global state? Remember that if the control structure added has a ring shape, the graph of the processes relative to the underlying calculation is *a priori* unspecified. Consider three processes placed on the ring in the following visit order: $(\ldots P_k, P_j, P_i, \ldots)$, at the instant when the token is at P_j: it has therefore visited P_k and observed it to be passive. P_i has not yet been visited, and if it is active it may (as a function of the underlying calculation and the corresponding linking) send a message to P_k (or to another process already visited by the token) and make it active (this activity may in its turn give rise to others). P_i can then become passive. When the token moves from P_j to P_i, then from P_i to its neighbour, the circulation rule ensures that it observes them to be passive before leaving.

It can be concluded from this counter-example that the observation by the token of all the processes as passive is not a sufficient condition allowing termination to be assumed (even with the assumption of empty channels). This can be explained by the fact that the activities of the processes are parallel whereas the observation by the token is sequential.

Solving the problem therefore requires that the token should be given a value, so that once it has gone all around the ring, the value will indicate whether the termination is effective or not. For this, the token is coloured with either of two colours, *yes* and *no*. If it is launched by P_0 with the colour *yes*, and returns to it with the same colour, the termination will be effective; if on its return, the colour is *no*, P_0 will be able to change its colour to *yes* again before relaunching a new detection.

The problem can be seen as consisting of developing a rule to allow a process P_k to colour the token with *no* if it has been able to activate a process P_i already visited by the token. This would be easy if each of the processes, knowing the order in which they are placed on the control ring, also knew at any instant the process in which the token is present (given the rendezvous hypothesis, it cannot be in a channel). During the sending of a message it would know if it had activated a process already visited. The first global knowledge (order of placing on the ring) is easy to achieve, either dynamically with a learning algorithm (cf. Chapter 3) or statically. The second solution is opted for here: the processes are placed on the ring in the order $P_0, P_{n-1}, P_{n-2}, \ldots P_2, P_1, P_0$, and each one knows this order. (If the processes do not have these identifiers in the underlying calculation, a preliminary tour by a control message can provide them.) The second element of global information (the location of the token) cannot be perceived by all the processes; a process P_i has only one of the three following elements of partial knowledge (given the ring structure): it has the token (it knows that $P_0, P_{n-1}, \ldots P_{i+1}$ have been visited); it has had it previously; or it has not yet had it (it knows that the token is in one of the processes $P_0, P_{n-1}, \ldots P_{i+1}$ but it does not know which). In the third case, P_i does not have exact (and therefore global) knowledge of which process the token is in and must consider the worst possibility from the point of view of the reactivation of processes observed to be passive, the one where the token is in P_{i+1}. The rule can therefore be stated: if P_i sends a message to a process with a control identifier greater than its own, it must colour the token with *no* when it passes it on. For this each process is equipped with an additional state variable $colour_i$; initialized to *white*, it is changed to *black* on transmission of a message to P_j if $j > i$. On transmission of the token, it is coloured *no* by P_i if $colour_i = black$.

The algorithm

The algorithm for P_i is finally as follows. The detection is initiated and observed by P_0 with the token moving in direction $P_0, P_{n-1}, \ldots P_1, P_0$.

```
on reception of message(m)
   do state_i := active end do
on waiting for message or end
   do state_i := passive end do
on transmission of message(m) to P_i
   do if i < j then colour_i := black end if
      send message(m) to P_j
   end do
on reception of token(coltoken) from P_(i+1) mod n
   do tokenpresent_i := true;
      if i = 0 then
         if colour_i = white and coltoken = yes
            then termination detected
            ifnot re-initiate detection
         end if
      end if
   end do
on transmission of token(coltoken) to P_(i+1) mod n
   possible only if state_i = passive and tokenpresent_i := false;
   do if colour_i = black then coltoken := no end if
      tokenpresent_i := false;
      send token (coltoken) to P_(i-1)
      colour_i := white
   end do
```

This control algorithm based on a ring structure is interesting, apart from the practical aspect, because of the methodological approach adopted for its formulation. As an exercise, the reader is recommended to examine the particular case where P_0 has initiated a detection, and a non-visited process P_i sends it a message: this means that $i > 0$ and P_i is not coloured *black*. Another exercise consists of moving the token in the opposite direction around the ring. Will the rules be the same?

Asynchronous communications case

The assumption of communication by pre-arranged rendezvous (that is, with empty channels) can be eliminated easily. Assume only that the channels are reliable—there is no desequencing. It is then necessary to visit the channels as well and observe them to be empty. To achieve this the ring is constructed as follows: on the oriented graph (assumed also to be strongly connected) defined by the underlying calculation communications, a cycle is constructed which includes each of the channels (at least once and possibly several times). The ring obtained therefore contains each of the processes at least once and each of the channels at least once. It can be used for the tour of the processes and channels. At the end of a circuit around the ring by the token, during which it finds the processes to be passive, nothing can be concluded concerning termination: messages in transit may have reactivated some processes. If, on the other hand, the token makes two circuits of the ring and in each observes the processes to be in the passive state, and if these processes have remained passive between the two circuits (simple boolean variables will establish this), then it can be concluded that all the processes are passive and the channels empty. This is based on the assumption of non-desequencing, which allows the period between two circuits to be defined as one in which the observation of the system is consistent from the point of view of detecting the stable property of termination.

Note: In these two algorithms (communication with and without rendezvous) a process must initiate the detection of termination in order to know whether the system is in the terminated state. This essential character of initiating the calculation differentiates these algorithms from those seen in Section 6.3.4; in this algorithm the process associated with the root of the control tree is informed of the termination by the change in the value of a predicate to *true*. These are two fundamental differences in the design of distributed algorithms which can be related to imperative and declarative styles. The same difference can be seen between the exclusion algorithm seen in Section 6.4.2.2, where each process asks its neighbour (in a ring) for the value of the state variable in order to test a condition, and that seen in Section 6.3.3, where each process systematically broadcasts (on a complete graph) the modifications in its state relative to the exclusion (request) and waits for a condition to be satisfied without ever making a request to the other processes.

6.4.2.4. THE PRODUCER–CONSUMER PROBLEM

Summary of the problem

In Chapter 1, the problem of the producer–consumer was introduced, and its solution in both reliable and unreliable contexts was studied (allowing the introduction of such tools as counters, sequence numbers, PAR protocols, etc). The production and consumption entities were reduced to a single process each. The case of a reliable context, in which the production entity is made up of *m* producer processes, is now studied.

The reader is referred to Section 1.3 for an implementation with two processes, but note the global synchronization constraints of the problem (expressed using the counters #*init*, #*term*; *n* is the number of buffers), summarized below:

$$condition(produce) = \#init(produce) - \#term(consume) < n;$$
$$condition(consume) = \#init(consume) - \#init(produce) < 0$$

It is assumed, as in Chapter 1, that the buffer is located at the consumer site, and only the distribution of control is to be considered (the results extend those in Chapter 5 on the distribution of global synchronization constraints).

A preliminary solution (non-fair)

The producer reads and writes the variable *startprod* which represents the abstract counter #*init*(*produce*): the *m* producer processes must therefore attain mutual exclusion if there is to be a consistent value. It was seen in Section 6.4.2.2 that a ring structure can help in achieving exclusion. The producers $P_1, \ldots P_m$ and the consumer *C* will therefore be placed in a ring as follows: $P_1, P_2, \ldots P_m, C, P_1$. This ring is only for control; once a producer has obtained the right to produce it will be able to communicate the message by the channel of its choice to the consumer, which will store it in the buffer. Since a process can only handle *startprod* in exclusion, a simple solution consists of valuing the mutual exclusion tag with *startprod*. The unique instance of this variable will then be accessible only in exclusion, and will always have a correct value. The condition associated with the producer P_i:

$$condition_i (produce) = startprod - cendcons < n$$

cannot be evaluated unless P_i has the token. (*ctermcons* is a local image at P_i with delayed updating of *endcons*.) The token can also transport the value *ctermcons*: it is used only in conjunction with *startprod*. When the token moves to process C, it can update it by *ctermcons* := *termcons* (in addition, the transport of *ctermcons* by the token eliminates the channels and the messages for broadcasting the value of *termcons* from *C* to each of the P_i).

Each process has a fair view of the token. When the token reaches it, a producer P_i that wants to produce tests the condition; if the value is *true* it increments *startprod*, passes the token on and sends its message to *C*; if it does not wish to produce it passes the token on directly. When it wishes to produce, and the condition does not authorize it, it must still pass the token on (if it did not do this the system would be deadlocked since *startprod* and *ctermcons* would never be modified). The condition therefore remains false until the consumer modifies *ctermcons*. It is assumed that each producer carries out only one production at a time and at the most one for each circuit of the token around the ring.

If there are more producers than buffers ($m > n$), the above protocol may lead to starvation of the producers in a queue around the ring. If the first n producers after the consumer want to produce, they can do so, if the condition is satisfied, and the condition is then always false for the $m - n$ other producers; these are therefore in a starvation situation.

The introduction of the above control guarantees each producer a fair view of the token (they all see it pass), but the use they can make of it is not fair.

A second (fair) solution

The above problems arise from the fact that when a process evaluates the condition (it then has the token) and finds it to be false, it does not know the variable *termcons* which would allow it either to produce or to wait until the variable reaches a value that would allow it to produce.

One solution consists of dissociating the handling of *startprod* (which must always be carried with exclusion between the producers) from the perception of *ctermcons*. The token transports only *startprod*, with the consumer C broadcasting the successive and increasing values of *termcons* to all the producers, which store them in local images *ctermcons$_i$*. This solution involves a great many messages, as already seen. When P_i has the token, it can test the condition; if this is false it retains the token (that is, the right to produce) until the local image *ctermcons$_i$* has reached a value authorizing it to produce. The processes that wish to produce, therefore, do so according to the order in which they are placed on the ring, thus ensuring fairness. The viewing and use of the token by the processes are here coordinated in such a way as to guarantee fairness.

A solution entirely on a ring

The aim is to find a solution that does not require another control topology apart from the ring.

The previous solution blocks the token at P_i until it is able to produce. If the token transports *startprod* and *ctermcons*, the moment at which a process sees the token pass must be dissociated from that at which it produces. To achieve this a new variable *numorder* is introduced, shared by the producers, which will allow them to be ranked among the production demands. The token is now made up of three values:

token (*startprod, ctermcons, numorder*)

In addition, each process P_i has a local variable *rank$_i$* in which it will store its order number.

When a producer P_i wishing to produce sees the token pass, it takes an order number.

numorder := *numorder* + 1;

rank$_i$:= *numorder*;

and the values transported by the token also provide it with the following information on the global state:

—the number of unsatisfied requests for production, the priorities of which are higher than the producer's request (that is, made previously and given priority to ensure fairness) is:

$$ndemprio_i = rank_i - (startprod + 1)$$

—the number of free places in the buffer as perceived by P_i is:

$$nfree_i \geq n - (startprod - ctermcons).$$

The production condition associated with P_i that guarantees fairness between the *m* producers is therefore that the perception of the global state by P_i assures it that there remains at least one free place when the priority demands have been satisfied:

$$condition_i \, (produce) = nfree_i - ndemprio_i > 0$$

This expression assumes that the producer P_i sets its rank before testing its local condition; in addition, $rank_i > startprod$ must be satisfied before it can produce.

From the local condition and from $rank_i > startprod$ a new formulation can be deduced for the local condition governing the production of P_i:

$$condition_i \, (produce) = rank_i - ctermcons \leq n$$

It will be noted that the local conditions of P_i satisfy the abstract and global production conditions:

$$\forall i : condition_i \, (produce) => startprod - termcons < n$$

After receiving the token and taking its rank, if the process P_i wishes to produce it tests its local condition; if the latter is satisfied, P_i increases *startprod* by 1, transmits the token and carries out production; in the contrary case, it transmits the token without modifying *startprod*. During subsequent transfers of the token, since the value of *ctermcons* has been modified by the consumer as a function of the messages received, it will test its local condition (note that it only produces messages one after the other; a request for production must have been honoured before another message is produced) and can produce once this is satisfied.

For a given P_i the expression $rank_i - ctermcons$ varies in a monotonically decreasing way during successive circuits of the token around the ring. In effect, $rank_i$ is constant and *ctermcons* is monotonically increasing: a producer therefore sees its requests for production accepted at the end of a finite period (the delay in transmission of the token is finite and time of usage is also finite).

The control algorithm ensures fairness and is based only on a ring structure. The fairness is linear: the waiting time of a producer is proportional to the number of producers. In the worst case, this wait is proportional to $n + m$: all the buffers are full and all the P_i wish to produce. This intuitive result can also be reached formally. From $\forall i : rank_i - startprod \leq m$ (less than m priority requests) and $startprod - ctermcons \leq n$ (system invariant) it is possible to conclude:

$$rank_i - ctermcons \leq n + m$$

In order to save on unnecessary circuits by the token, it can be envisaged that the consumer should wait before retransmitting the token if nothing has been consumed since the token's last passing. A producer will wait at the most m circuits of the token around the ring. Note that if the processes that can produce transmit the token before producing the order of production is not necessarily the order of the requests for production; and both these orders can be different from the order of the messages' arrival at the consumer (which depends on the transfer time of the messages).

This problem has permitted the examination of the effect of ring structure on control algorithms—of what it contributes and what has to be added. The same problem could have been treated in another context—that of a complete graph, for example. The variable *startprod* could then be duplicated at each site but an algorithm for excluding these various copies must be provided to guarantee consistency (cf. Section 6.4.4.2).

6.4.3. The tree structure

6.4.3.1. GENERAL REMARKS

The tree structure is a fundamental structure (of data or control) in sequential calculation. It is found in distributed control as a structure of privileged communication in many algorithms. In the case where such a structure must be established dynamically, traversal algorithms can be written, as seen in Section 3.3. But the structure could also be defined statically.

The process associated with the root generally plays a special role. The broadcasting of information by this process to all the others is simple, as is broadcasting with general acknowledgement: a process does not send the acknowledgement to its parent in the tree until it has received all the acknowledgements from its sons.

The counting principle found in the protocol for broadcasting–acknowledgement is the basis of many distributed algorithms. In the algorithm for detection of termination seen in Section 6.3.4., for example, a process is placed in the control tree when it becomes active and is not already there, and cannot leave until it becomes passive again and has no sons, that is, it is not at the origin of any activity still continuing.

A distributed algorithm will now be examined, based on a statically defined tree structure; to vary our treatment, we shall study distributed algorithms for sorting and ranking.

6.4.3.2. A RANKING ALGORITHM

The problem of ranking

Consider n processes with different identifiers $id(1), \ldots id(n)$. Associated with each of these processes is a value (not necessarily distinct): $v(i)$ is the value held by the process with identifier $id(i)$. The ranking consists of arranging the processes in increasing order of the values $v(i)$. This involves giving them new identifiers $id'(i)$ between 1 and n, all different and such that:

$$v(i) < v(j) => id'(i) < id'(j)$$

Process structures

The whole process is considered as having a tree structure. Each of the processes has a working context: the multiset s_i associated with the identifier node $id(i)$. (A multiset is a set in which an element may appear more than once.)

Principle of the algorithm

The distributed algorithm is very simple. It is made up of a starting-up phase and a processing phase. The principle is a bubble sort applied to a heap structure (heapsort): this involves moving the smallest value from the leaves to the root, thus associating a rank order with the process from which the value comes, then reiterating until all processes have been numbered. The principle is therefore the use of the hierarchy provided by the tree structure.

Starting-up phase

The multiset of $id(i)$ is initialized with $v(i)$. Initially the root broadcasts a request to its sons, who report the process until the leaves are reached. A leaf process with identifier $id(i)$ then sends its parent its value $v(i)$ and removes this value from its multiset; on reception its parent $id(j)$ places it in its own multiset s_j. When it has obtained a value from each of its sons a node process with identifier $id(i)$ selects the smallest—let it be v—removes it from its multiset and sends the pair $(v, id(i))$ to its parent. When the root has obtained a value from each of its sons it knows the smallest value and through which of its sons it came. The starting-up phase is then ended.

Processing phase

The root then eliminates the smallest value v from its multiset and sends a message containing the ranking c (initialized to 0), having incremented it by 1, to the son through which the value v reached it. The message containing c is transmitted (each process knows the issuer of the last value it has sent to its parent) closer and closer to the process with which it was associated v and finally reaches it. This is numbered with the value of c and removes the smallest value from its multiset, sending it to its parent. If the multiset is empty an empty message is sent. The parent therefore obeys the above rules after having received a value (or *empty*) it adds this to its multiset and removes the smallest value, which it sends to its parent, etc.

Note: This algorithm is a type of distributed bubble sort, as already explained. Its operation is simple, and the reader is recommended to work out a structured algorithmic formulation for it, as an exercise.

The starting-up phase requires $O(n)$ messages: the algorithm global complexity is $O(n^2)$ in the worst case. It is easy to show that each value $v(i)$ exists only in one multiset and that a node does not send *empty* to its parent until all the processes of its subtree have been ranked.

This algorithm can be used to sort n values, each being associated with a process. An order of exit or value ranking can be added to the root as it eliminates them from its multiset.

6.4.3.3. A SORTING ALGORITHM

The sort problem

As before, n processes are considered with the identifiers $id(1), \ldots id(n)$, all distinct, on which there is a relation of total ordering. Associated with the process with identifier $id(i)$ is the value $v(i)$. Sorting the values $v(1), \ldots v(n)$ therefore consists of finding a permutation v' of the values such that:

$$id(i) < id(j) => v'(i) < v'(j)$$

This involves moving the values $v(i)$ between the processes so that at the end of the algorithm, the ordering of the values placed in the processes is the same as the ordering of their identifiers.

Principle of the algorithm

The sort principle is simple and, as before, is based on the hierarchy created by the communication tree-structure: it consists of extending the ranking algorithm by making it work both on the values $v(i)$ and on the identifier $id(j)$, then synchronizing these two parts and finally composing the results. When the root has obtained, at a given moment, the smallest value x and the smallest identifier y, it modifies the process y, not with an order of ranking, but with the value x.

For this, each process $id(i)$ has two local variables: a multiset s_i which will contain the values and a set r_i which will contain the identifiers: these variables are respectively initialized to $v(i)$ and $id(i)$. The spatial complexity of each process will be $O(n)$ as in the ranking algorithm.

In the starting-up phase, the root sends requests that are transported to the leaves. A leaf $id(i)$ then removes the value contained in its multiset s_i and the identifier contained in r_i and sends them to its parent. The parent $id(j)$, when it receives them, places them respectively in s_j and r_j. When a node has received couples of this sort from all its sons, it removes the smallest elements from s_j and r_j

and sends them to its parent. The starting-up phase is over when the root has obtained replies from each of its sons.

The processing phase can then start. The root $id(rt)$ knows the smallest value x and the smallest identifier y, and removes them from s_{rt} and r_{rt} respectively, sending a message containing the value x to the process with identifier y and a message requesting a new value to the process that had x as its initial value. As in the ranking algorithm, a node knows which of its sons sent it a value or identifier; a message transmitted by the root can therefore reach its destination. The node $y = id(k)$ then considers the value x as its final value, removes the smallest element from r_k and transmits it to its father (or the value *empty* if $r_k = \emptyset$). The process that had x as its initial value removes the smallest element from s_k and also sends it to its parent (it sends *empty* if $s_k = \emptyset$). The protocols for transmitting a value or identifier to the root are the same as the one used in the ranking algorithm. The algorithm stops when $s_{rt} = r_{rt} = \emptyset$.

This algorithm requires $O(n^2)$ messages to sort n values. It is very interesting from the methodological point of view because it works by synchronizing two classification algorithms that operate respectively on the values and the identifiers. The principle that can be derived from these algorithms based on tree structures is that each node of the tree 'controls' the subtree of which it is the root and has a global state for it (as regards this problem, the state is the smallest value and/or identifier present in the subtree). When it discovers that the perception it has of this state is not up to date, a node uses a protocol to update it: this protocol starts up the transfer of messages from a node possibly to the leaves and back to the node.

Other types of sort

The previous algorithm assumes that there is one process per value to be sorted; this is not always possible and in the general case there are $n(i)$ distinct values associated with the process with identifier $id(i)$. Performing out a sort then consists of storing all the values in order and placing them in the various processes.

The expression 'static sort' is used when at the end of the sort each process has the same number of values as it had initially (the above sort is static). The expression 'dynamic sort' refers to the contrary case. Dynamic sorts are useful when, for example, the memory occupation of the processors is to be balanced and the distribution of initial values is not.

These distributed sorting algorithms are very useful in systems for distributed file management; they allow the classification, ordering and merging of distributed files according to the access keys of their elements.

6.4.4. The complete graph

6.4.4.1. GENERAL REMARKS

When the graph is complete all the processes have some *a priori* knowledge: the number of processes and the graph itself are two examples. In a ring structure, however, the same does not apply: a process has prior knowledge only of the structure and not of the total number of processes. There are also other elements of *a priori* knowledge which are a function of the algorithm realized on a complete graph. If the requests of each of these processes are systematically broadcast, a process that receives a request knows that it has been or will be received by each of the sites (this assumes the delay in the transmission of messages to be arbitrary and finite): the situation is quite different in the case of an arbitrary graph where, if each request must be known by all, each process must, on reception of a message, propagate it to all its neighbours, apart from the issuer (see Chapter 3).

As was also explained in Chapter 3, the complete graph (and more generally all the *a priori* knowledge of the processes) presents a major disadvantage: its modification (suppression or addition of new processes) requires the redefinition of elements of each of the processes and therefore a recompilation of each of them. A compromise must be found between resilience to change (recovery after breakdowns, for example) and the global knowledge with which the processes are implicitly equipped. This will depend on the problem in question.

In this part, the distribution of a state machine in a complete graph will be considered. The example chosen as illustration is that of mutual exclusion but the principle of distribution is general (cf. Section 6.3.5).

6.4.4.2. DISTRIBUTING A STATE MACHINE

An allocation problem

Consider a resource of which there is only one copy, which for consistency must be accessed only in mutual exclusion. Two operations are defined to this end: *require* and *free*.

A state machine carries out the control of access to the resource as follows. From the external point of view, it accepts the operations issued by the processes one by one. From the internal point of view, it manages a queue (fifo) in which it places the requests as they arrive. It authorizes the process whose request is at the head of the queue to use the resource. The operation *free*, issued by the process holding the resource, cancels its request (which is at the head of the queue), which has the effect of assigning the resource to the process whose request is then at the head (if the queue is not empty).

This type of state machine can easily be implemented in the context of centralized parallelism using synchronization tools provided either by languages (semaphores, monitors, tasks, files and conditions) or by machines. The method for distributing it will be explained in the context of a distributed system.

The state machine chosen to illustrate this method (allocator of a resource used in exclusion) can easily be extended to the solution of other problems—the consistency of copies in a set of shared variables, for example (it can thus be applied to ensure the consistency of the copies *startprod*$_i$ of the variable *startprod* in Section 6.4.2.4 when the producers P_i are connected by a complete graph). More generally the problem is one of the consistency in updating of files with multiple copies.

Principle of distribution

The idea consists of providing each of the sites/processes P_i with a local state machine a_i and protocols between these machines such that the site shows the same behaviour as the global state machine A (which is abstract).

In a manner similar to the duplication of data on various sites (with suitable control to ensure consistency), one solution consists of duplicating A on each of the processes P_i and ensuring that the resulting local machines a_i all behave in the same way. As indicated in Section 6.3.5, it is necessary that all the machines a_i should perceive the same events in the same order. Since they are identical, start in the same initial state and use the same input sequence, they will all behave like A using the corresponding input sequence.

This distribution of A requires two things:

—the inputs (events) of which P_i is the origin must be broadcast to all the machines a_j (they perceive the same events);

—whatever their origins the inputs must be totally ordered (the machines perceive the events in the same order).

Two tools are therefore necessary; broadcasting, and a mechanism allowing the inputs (events: resource request, right of use, end of use by the process) to be identified singly so as to order them totally. Broadcasting is easy to achieve using a complete graph, and identification is easy using a system of logical clocks (cf. Section 4.2.2). It was seen in this chapter (Section 6.2.2) that an event can be identified in a unique way by the pair formed by the value of the local process clock (at the moment when the event occurs) and the identifier of this process. Note the total ordering relationship: since e_i and e_j are two events temporally identified by (h, i) and (k, j), then:

$$(e_i \ \textit{precedes} \ e_j) \ \textit{iff} \ (h < k \ \textit{or} \ (h = k \ \textit{and} \ i < j))$$

The logical clocks and the identifiers of the processes/sites allow a total ordering to be constructed. There remains a problem to be solved concerning the perception by the state machines of this ordering of the events. Consider the machine a_i and event e_j (identified by (h, j)) which it has perceived. How does a_i know that e_j is the next event to be processed? It must be the oldest in the sense of the total ordering defined above, that is whatever the event e_l identified by (k, l) to be processed, e_l *precedes* e_j must apply, that is:

$$h < k \ \textit{or} \ (h = k \ \textit{and} \ j < l)$$

If a_i has perceived an event e_l such that e_l *precedes* e_j, then it knows that e_j is not the next event to be processed. In the same way, if a_i has perceived the events e_1, $e_2, \ldots e_n$ coming from each of the sites other than P_j (from which it has received e_j), and they have not yet been processed, it knows that e_j is the next event to be processed if:

$$\forall l \neq j : e_j \text{ } precedes \text{ } e_l$$

On the other hand, what can it conclude if it has perceived only the events e_l from certain site and for each of them e_j *precedes* e_l? It cannot conclude that it can process e_j immediately: in fact it may next perceive from one of the sites an event (via a message) e_m such that e_m *precedes* e_j. A machine a_i therefore cannot conclude that an event is the next to be processed unless it has perceived an event from each of the sites: the one it processes will be the oldest, in the sense of the total ordering defined.

If a site does not produce an event the system will be deadlocked: no machine has all the elements allowing it to order the events. To overcome this problem, the local machine a_i must be offered means of finding out about the absence of events; inquiry messages are used for this.

This is a standard global state problem that has been encountered previously: a machine a_i must ensure, before taking a decision, that the view of the global state on which the decision will be based is consistent and will never be challenged (cf. Sections 6.3.2 and 6.3.4).

The events local to P_i are perceived directly by a_i and broadcast with their identifiers to the other sites P_j using messages. For the perceptions of the P_j to be consistent the messages must not be lost, overtake each other or be desequenced.

Distribution of the allocator

Each machine a_i has the same declarations of variables allowing it to detect that pair of the global state part which concerns the problem to be solved. The protocols that link the behaviour of the machines a_i use messages; which here are of the following type:

type typmess = (*request, ack, freeing*)

The *request* or *freeing*-type messages are used by the protocols that broadcast the events; requests for and freeing of the resources by the processes.

When a process P_i wishes to use the resource, machine a_i broadcasts a message request(h_i, i). When a controller a_j receives such a message it may resynchronize its clock h_j and reply to a_i with the message $ack(h_j, j)$. This 'call–reply'-type protocol systematically responds to the enquiry that would be initiated by a_i if a_j did not broadcast a request allowing a_i to conclude (cf. the previous discussion). Note that the later requests of a_j will be identified with dates later than that which accompanies the request received (because of the systematic resynchronizing of clocks during message reception).

In addition to a clock h_i, each machine a_i is equipped with an array that implements the queue locally:

var $queue_i$: **array** $[1 .. n]$ **of** (*typemess, identity*)

The field identity is made up of a clock value and a site identifier: the element $queue_i[j]$ denotes the last event known from a_j, with its temporal identifier. Since the resource is initially free:

$\forall i, j$: $queue_i [j] = (freeing, (0, j))$

Machine a_i authorizes P_i to use the resource only when the corresponding request is the oldest event in $queue_i$ (in the sense of the ordering relationship on the events' identifiers). When the resource is given to P_i, no other request can have a request identifier smaller than its own: it is therefore the only one able to use it because the machines a_i perceive the same events in the same order, but not necessarily at the same moment.

The behaviour of the control machines a_i on reception of a *freeing* or *ack* message from a_j remains to be defined. As far as the former is concerned, a_i resynchronizes h_i and updates $queue_i [j]$ with the message. As far as the latter is concerned, a_i also resynchronizes h_i but does not update $queue_i [j]$ unless this variable does not contain (*request*, . . *j*): in fact this type of value can only be cancelled by a message (*freeing*, . , *j*).

In the following text, the test on the request identifier fields is defined on the integer pairs: this is the test derived from the total ordering relationship on the identifiers.

```
on a call to request
    do h_i := h_i + 1;
        queue_i [j] := (request, h, i);
        broadcast request (h_i, i);
        waitfor ∀ j ≠ i : queue_i [i], identity < queue_i [j], identity;
        - - the wait can be interrupted
    end do
on a call to free
    do h_i := h_i + 1;
        queue_i [i] := (freeing, h_i, i);
        broadcast freeing (h_i, i)
    end do
on reception of request (k, j)
    do h_i := max(h_i, k) + 1;
        queue_i [j] := (request, (k, j));
        send ack(h_i, i);
    end do
on reception of freeing (k, j)
    do h_i := max(h_i, k) + 1;
        queue_i [j] := (freeing, (k, j));
    end do
```

on reception of *ack* (k, j)
  do $h_i := max(h_i, k) + 1;$
    if *file*$_i$ $[j] \neq (request, . , j)$
      then *queue*$_i$ $[j] := (ack, (k, j))$
    end if
  end do

This algorithm requires $3(n-1)$ messages for each use of the resource. The order of requests for the resource is the same as that in which they are satisfied: this is the total order of their identifiers (temporal and topological). Since the requests for a single process are made one after the other and all the demands are ordered, there is no starvation. The construction and use of a total ordering (static on the resource identifiers or as here, dynamic, on the request identifiers) constitute a general method for avoiding starvation situations.

6.5. Knowledge of neighbours

6.5.1. Introduction

In Chapter 3 basic algorithms for network learning and traversal were introduced, and the assumption was made that initially the processes do not know their neighbours. What would be the effect of the contrary assumption? Would *a priori* knowledge of this information be an important tool for the design of distributed algorithms?

Consider, as an example, the problem of broadcasting information from one process to all the others. Apart from its identifier and those of its neighbours, a process (site) knows nothing about the structure of the network (apart from its connectivity) or the total number of processes. (The reader is referred to Section 3.3.2, where an algorithm for parallel traversal is given, allowing, among other things, broadcasting to be achieved). Let v be the value that P_r must broadcast, *neighbours*$_r$ the set of identifiers of its neighbours and x one of these identifiers.

Knowing the identifiers of the processes to which it will send the value v, the process P_r can construct messages made up of two fields: a data field and a control field. Into the former it places the value v and into the latter the identifiers of the processes that will have knowledge of the message, that is $\{r\} \cup$ *neighbours*$_r$. The value that must be broadcast is therefore 'enveloped' in the message:

broadcast $(v, \{r\} \cup$ *neighbours*$_r)$

Now consider a process P_x which is a neighbour of P_r. On reception of such a message, which takes the form *broadcast* (v, z), P_x knows that all the processes whose identifiers are included in z have received or will receive the same message: there is therefore no point continuing the broadcast of v to them if they are neighbours of P_x. Generally speaking, a process P_i obeys the following behaviour (the variables used are self-explanatory):

on reception of broadcast (v, z) *from* P_j
   do
      if ¬ *received$_i$ then*
         received$_i$:= true:
         dest$_i$:= neighbours$_i$ $- z : --$ so $j \in z$
         ∀ $x \in$ *dest$_i$* : *send broadcast* $(v, z \cup$ *neighbours$_i$) to* P_x
      end if
   end do

The control knowledge transported in z therefore allows the number of messages exchanged to be reduced relative to the algorithms that do not employ this information. The messages that carry out the broadcasting of v have values of z that depend on the speeds of the messages through the network and on the specific configuration of the network. At maximum, z contains n identifiers; the added complexity in the size of the message is therefore proportional to the number of processes.

This simple process can be applied to many algorithms. More generally, and as a function of the problem to be solved, a process may, on reception of a message with a data field v and a control field z, modify them both before moving them on:

on reception of message (v, z)
   do
      calculations on v that give v′:
      calculations on z that give z′:
      transmit : *message* (v', z'):
   end do

In the following section, ways of using *a priori* knowledge of the identifiers of neighbours to achieve an election in a network will be studied.

6.5.2. An election algorithm

The problem of election

Consider an arbitrary (connected) network in which the channels are bi-directional and reliable each process knows only its neighbours by their iden tifiers. During the calculation, a process must not learn global information (such as the structure of the network, for example) that it could use later. This constraint eases the problem of recovery after breakdowns.

A distributed algorithm that can be started up by one or more processes must be designed such that at the end of the algorithm a single process has been elected by the others, and such that each of the others knows its identifier, and which of its neighbours allows the elected process to be reached. Since any process could, *a priori*, be elected (it does not play any special role until after the election) the one with the largest (or smallest) identifier is usually chosen for election, which assumes that the identifiers are all distinct and totally ordered. These are the only assumptions made concerning the identifiers.

Solution principle

The problem can be reformulated as follows: a tree must be constructed, the root of which is the process with the largest identifier; at the end of the algorithm each of the processes must know the root, which of its neighbours allows it to be reached (i.e., which is in the tree) and which of its neighbours are its sons in the tree.

One of the most difficult problems arises from the requirement that the algorithm can be started by any number of processes.

The principle on which the algorithm will be constructed is the following. For a process P_i, initiating an election consists of starting an exploration of the network the aim of which is to visit all the processes and establish routes from the processes to itself. As the algorithm can be started by several processes, several explorations may take place simultaneously. At the end of the algorithm, the exploration started by the process with the largest identifier must alone have terminated correctly (that is, visited all the processes) and this must be the last exploration taken into account by the other processes for the establishment of the tree. An exploration can be identified uniquely by the identifier of the process by which it was begun. The two above objectives will be attained when: any process P_i visited by an exploration of weight j stops it without taking it into account, either if P_i has been visited previously by an exploration with identifier k, with $k >$ j, or if it has never been visited and $i > j$ in the latter case P_i initiates an exploration of identifier i, if it has not already done so. In all other cases it takes account of the exploration of weight j and passes it on.

Two types of exploration issuing from a process P_i can be envisaged: in parallel (the progression is towards all the neighbours) or depth-first (the progression is towards each neighbour in turn).

Chapter 3 contained details of parallel and depth-first traversal. The principle of the former was used in the previous section (6.5.1) to illustrate broadcasting using the field z, namely an understanding of control based on neighbours' identifiers. In this chapter, explorations by depth-first traversal will be used. The reader is recommended to re-read Section 3.3.1 and could, by way of an exercise, formulate an election based on explorations by parallel traversal.

Using neighbours' identifiers

Depth-first exploration is carried out with messages of the type *explore*. Such a message will transport three items of information: k, the identifier of the process that initiated the corresponding exploration; z, the set of identifiers of the processes visited by this message; and s, the set of identifiers of those neighbouring the processes of z that have not yet been reached by this exploration. During its progress, the message *explore* (k, z, s) examines its fields z and s being modified, and while $s \neq 0$ the exploration is not ended. (In terms of graphs, s represents the vertices not yet visited.) The stopping condition of the exploration with the maximum identifier is applied to a field destined to contain the identifiers (the other explorations will have been stopped previously, cf. the stated principle).

Since the exploration takes place with a depth-first traversal, the messages *backtrack*(k, z, s) are introduced. They allow an exploration that has not reached all the processes ($s \neq 0$) and can progress (since it has not yet been stopped by a process), but is not able to do so from the process currently visited, to start off again from a process already visited that has a neighbour belonging to s.

Relative to the message *scan*(k) and *backtrack*(k) of a non-informed depth-first traversal (Section 3.3.1) it can be seen that explicit knowledge of the neighbours leads to a saving in the number of messages and amplifies the design of the traversal algorithm.

The election algorithm

Once the principles of the election algorithm (progression and stopping of exploration) are established and a model of the informed depth-first traversal has been defined to carry out the exploration, the algorithm assembles all the components.

If the processes must be informed of the termination of election it is necessary to introduce a message of type *conclude* which is initiated by the process that finds $s = 0$ and indicates to the other processes that the election is terminated and that consequently the variables indicating the largest identifier, their parent and their sons in the tree, have their final values.

Each process P_i is given the following context:

const neighbours$_i$ initialized to (*identifiers of neighbours of P_i*);
var state$_i$: (*initial, candidate, beaten, elected*) *initialized to initial*;

This variable gives the current state of P_i; at the end of the algorithm only one of these variables has the value *elected*, and the others have the value *beaten*.

var lgstseen$_i$: *integer initialized to* i:

This concerns the largest identifier that P_i has seen pass; at the end of the algorithm it will contain the largest of the identifiers.

var parent$_i$: *integer initialized to nil*;
  son$_i$: *set of integers initialized to* 0;

These two variables place P_i in the tree relating to the exploration of the largest identifier to have visited P_i.

To make the behaviour of P_i easier to express, the following abbreviation is introduced:

procedure initiate exploration;
        begin
            let $x =$ *maximum* (*neighbours$_i$*);
            state$_i$:= *candidate*;

$parent_i := nil;$
$son_i := (x);$
send *explore* $(i, \{i\}, neighbours_i - \{x\})$ **to** P_x
end

For the depth-first traversal we choose to move an exploration towards the neighbour with the largest identifier: other heuristics can be envisaged. (The heuristic chosen involves stopping an exploration as early as possible: for this purpose the exploration is moved towards the largest neighbours).

The behaviour of P_i is as follows:

on decision to initiate an election
   do
      if $state_i = initial$ **then** *initiate exploration* **end if**
   end do
on reception of *explore* (k, z, s) **from** P_j
   do
      case $lgstseen_i > k \rightarrow$ **if** $state_i = initial$ **then** *initiate exploration* **end if**
           $lgstseen_i < k \rightarrow state_i := beaten;$
                 $lgstseen_i := k;$
                 $parent_i := j;$
                 let $y = neighbours_i - z;$
                 case $y = \varnothing \rightarrow sons_i := \varnothing;$
                         case $s = \varnothing \rightarrow$ **send** *conclude* **to** P_j
                             $s \neq \varnothing \rightarrow$ **send** *backtrack* $(k, z$
                                        $\cup \{i\}, s)$ **to** P_j
                         end case
                    $y \neq \varnothing \rightarrow$ **let** $x = maximum(y);$
                            $son_i := \{x\} ;$
                            send $explore(k, z \cup \{i\}, s \cup y - \{x\})$ **to**
                                                 P_x
               end case
     end case
   end do
on reception of *backtrack* (k, z, s) **from** P_j
   do
     if $lgstseen_i = k$ **then**
          let $y = neighbours_i \cap s;$
            case $y = \varnothing \rightarrow$ **send** *backtrack* (k, z, s) **to** P_{parent_i}
                   $y \neq \varnothing \rightarrow$ **let** $x = maximum(y);$
                         $sons_i := sons_i \cup \{x\};$
                         send $explore(k, z, s - \{x\})$ **to** P_x
          end case
     end if
   end do
on reception of *conclude* **from** P_j

> *do*
> *if lgstseen$_i$ = i then state$_i$:= elected* **end** *if;*
> $\forall\, x \in (son_i \cup parent_i) - \{j\}$: **send** *conclude* **to** P_x
> *end do*

This algorithm requires $0(n^2)$ messages. The best election algorithms operate on networks whose topology's either a complete graph or a ring and the processes have this global information; they have a complexity of $0(n \log n))$.

In connection with distributed algorithms, the algorithm shown is interesting for two reasons: firstly, because of the global knowledge that the processes have, consisting of the identifiers of their direct neighbours; secondly, because of the use made of this knowledge. The fields z and s of the messages avoid totally blind traversal having to be made, and the field s allows the end of the traversal, relative to the largest identifier, to be detected as soon as possible. Note also that in this distributed context, s plays the same role as the stack in a sequential context.

6.6. References

The methodology for design and writing of algorithms and programs is a subject that has been thoroughly studied in a sequential context (Dahl *et al.* [1972], Dijkstra [1976], Gries [1981]); the subject has also been tackled in the field of parallel programming (Hoare [1985], Brinch Hansen [1977], Schiper [1986], Filman and Friedman [1984]). This list of references is, of course, not exhaustive.

As far as the temporal aspect is concerned, the algorithm for the detection of deadlock based on the bound Δ used as an illustration is due to Wuu and Bernstein (1985). Helary *et al.* (1986a) introduce an algorithm for detection of deadlock, linked not to use of resources but to communications between processes, also based on prior knowledge of Δ. This knowledge allows the concept of an observation period to be defined, at the end of which the global state observed is consistent; Chandy *et al.* (1983) and Chandy and Misra (1985) contain details of this concept and apply it to the detection of deadlock due to communications and to distributed termination. Deadlock due to resources constitutes a problem that has been studied in depth. Merlin and Schweitzer (1980), Gunther (1981) and Gopal (1985) give solutions to this problem when it is due to the use of buffers in store-and-forward networks. These solutions are based on techniques of static ordering of resources, found in centralized systems (Havender [1968]). The principle of dynamic ordering of requests constitutes a technique for avoiding starvation in a distributed context: this is the principle applied in Section 6.4.4.2 to guarantee fairness between requests when distributing a state machine. This principle was introduced by Lamport (1974) and consists of associating with every request an order number and the identifier of the site that produces it. It has been used in many algorithms where the graph is either complete or arbitrary; for example, Ricart and Agrawala(1981) use it to solve mutual exclusion. Schneider *et al.* (1984) use it for reliable broadcasting and Chandy *et al.* (1983) use it to avoid deadlock.

The exclusion algorithm introduced in Section 6.3.3 was devised by Ricart and Agrawala(1983) or by Suzuki and Kasami(1982), both of which papers present effectively the same algorithm! For further information on mutual exclusion, see Raynal(1985). The algorithm for detection of termination in Section 6.3.4 was devised to demonstrate the importance of Δ, and was originally inspired by Dijkstra and Scholten(1980), in which transfer delays are finite but the necessary state variables are *a priori* unbounded, as is not the case in the algorithm given.

The assumption of a bound Δ occurs in many algorithms. Consider particularly algorithms for election in complete networks—the 'bully' algorithm—(Garcia Molina [1983]) algorithms that tackle the problem of the distributed consensus, more generally called the Byzantine Generals problem (Lamport *et al.* [1982], Dolev [1982], Dolev and Strong(1983), Fischer *et al.* [1985], Perry and Toueg [1986]).

The distribution of state machines is an important problem in the design of distributed systems; Lamport makes an excellent study of it. Lamport (1978) introduces the algorithm given in Section 6.4.4.2. Lamport(1984b) and Lamport and Schneider (1985) constitute an exhaustive study of the distribution of state machines in an unreliable environment for which the bound Δ is assumed. Note that although messages can be lost, their bound is not too restrictive: messages whose times of transit are greater than Δ are eliminated. This type of behaviour appears in the solutions to the Byzantine Generals problem; to solve the problem, the number of faults—breakdowns, loss, etc—must be less than a given number, which is a function of the number of processes.

As concerns the structural aspect, many more studies have been made. The mutual exclusion algorithm on a ring given in Section 6.4.2.2 is by Raynal (1985b). There are several in Martin(1985); Verjus and Thoraval(1986) study ring structure in relation to the problems of arbitration and of allocation of resources. When a token is used to represent a privilege on a ring, the problem of its loss arises if the lines are not reliable; Le Lann(1977), Misra(1983) and Raynal and Rubino (1985) provide algorithms for detection of loss, and regeneration of the token.

The termination algorithm introduced in Section 6.4.2.3 is by Dijkstra *et al.* (1983). The termination algorithm also based on a control ring but without the assumption of rendezvous is by Misra(1983); the two circuits made by the detection token represent a period of observation. Apt and Richier (1985), includes other algorithms for termination on a ring.

Rings have been thoroughly studied, particularly in the case of election algorithms (Chang and Roberts [1979], Dolev *et al.* [1982], Franklin [1982], Hirschberg and Sinclair [1980], Peterson [1982], Pache *et al.* [1984]). They are described, analysed and compared in Raynal (1985).

The problem of the producer and consumer on a control ring is studied in Plouzeau *et al.* (1987); other solutions to the problem are also given, notably concerning the distribution of a state machine in a complete graph.

Kumar (1985) presents an interesting study on the termination of distributed calculation; the solutions are based on ring structures and the use of markers and counters. Different types of termination detection algorithms are defined as a function of the tools used.

The tree structures used to represent control in networks and distributed systems have been thoroughly studied (Dijkstra and Scholten [1980], Segall [1983], Chandy *et al.* [1983], Cheung [1983], Chandy *et al.* [1983], Awerbuch [1985], Helary *et al.* [1986a]). Schneider *et al.* (1984) present an interesting application for reliable broadcasting in a context where the processes break down and stop completely (fail-stop processor). The algorithms given for classification and sorting introduced here are by Zacks (1985); the concepts of static and dynamic distributed sorting have been studied by Rotem *et al.* (1985). An original use of the tree structure to achieve mutual exclusion is presented by Trehel and Naimi (1986): a tree that restructures itself dynamically, and whose processes are nodes, acts as the structure for controlling the exclusion.

The assumption of knowledge of the identity of neighbours has been studied in particular, in the context of arbitrary networks, by Helary, Maddi and Raynal. They have applied it particularly to the detection of deadlock (1986a), mutual exclusion (Helary, Plouzeau and Raynal [1986c]) and the problem of election—the algorithm given in Section 6.5.2 is from Helary, Maddi and Raynal (1986b).

The broadcast primitive also constitutes an important tool for writing distributed systems, but has not been discussed here. The interested reader is referred to Gehani (1984) for the methodology and to Dechter and Kleinrock (1986) and Babaoglu and Drummond (1985), which discuss the problems of sorting and of consensus respectively, using broadcast structures.

Specific regular structures other than the ring, tree or complete graph can also be envisaged. Such are, for example, hypercube-type structures Seitz (1985) or communication structures based on projective planes Lakshman and Agrawala (1986). Maekawa (1985) gives an algorithm for mutual exclusion in which, although the graph is complete, the sites are grouped into non-disjoint subsets satisfying certain properties in Lakshman and Agrawala (1986).

Readers interested in distributed algorithms and protocols for specific functions or control in standard graphs will find in Raynal (1985a) an introduction to the problems in a distributed context and the algorithms for their solution.

Bibliography

AHO A.V., HOPCROFT J.E., ULLMANN J.D. (1974): *The Design and Analysis of Computer Algorithms*; Addison-Wesley, 470pp.

ALPERN B., SCHNEIDER F.B. (1985): Defining liveness; *Inf. Processing Letters*, Vol. **21**, pp. 181–185.

ANDRE F., HERMAN D., VERJUS J.P. (1983): *Synchronisation de programmes parallèles*; Dunod, 138pp.

APT K. (1986): *Correctness Proofs of Distributed Termination Algorithms*; ACM Toplas, forthcoming.

APT K., RICHIER J.L. (1985): *Real Time Clocks versus Virtual Clocks*; NATO ASI Series, Vol. **14**, Springer-Verlag, pp. 475–501.

AWERBUCH B., (1985): A new distributed depth first search algorithm; *Inf. Processing Letters*, Vol. **20**, pp. 147–150.

BABAOGLU O., DRUMMOND R. (1985): Streets of Byzantium: networks architectures for fast reliable broadcasts; *IEEE Trans. on Soft. Eng.,* Vol. **SE 11**, 6, pp. 546–554.

BARTLETT K.A., SCANTLEBURY R.A. WILKINSON P.T. (1969): A note on reliable full-duplex transmission over half-duplex links; *Comm. ACM.*, Vol. **12**, 5, pp. 260–261, 265.

BEN ARI M. (1982): *Prentice Principles of Concurrent Programming*; Prentice Hall, 172pp.

BERNSTEIN P.A., GOODMAN N. (1981): Concurrency control in distributed database systems; *ACM Computing Surveys*, Vol. **13**, 2, pp.185–201.

BERNSTEIN P.A., GOODMAN N. (1982): A sophisticate's introduction to distributed database concurrency control; *Proc. of the 8th Int. conf. on VLDB*, Mexico, pp. 62–76.

BOCHMANN G.V. (1979): Distributed synchronization and regularity; *Computer Networks*, Vol. **3**, pp. 36–43.

BOCHMANN, G.V. (1983): *Concepts for Distributed System Design*; Springer Verlag, 259pp.

BOKSENBAUM C., MUNTEAN T. (1976): Synchronization by means of constraints: formulation and applications; *Proc. 2d Symposium on Programming*, Paris, (Dunod Ed.), pp. 9–18.

BOUGE L. (1985): *Symmetry and Genericity for CSP Distributed Programs*; Rapport de Recherche, LITP, Paris, 22 pp.

BRINCH HANSEN P. (1977): *The Architecture of Concurrent Programs*; Prentice Hall, 317 pp.

CARVALHO O. (1985): *Une Contribution à la Programmation des Systèmes Distribués*. Thèse d'Etat, Université de Paris-Orsay.

CARVALHO O., ROUCAIROL G. (1982): On the distribution of an assertion, *Proc. of ACM-SIGOPS Symposium on PODC*, Ottawa, pp. 121–131.

CARVALHO O., ROUCAIROL G. (1983): Assertion, decomposition and partial correctness of distributed control algorithms. In *Distributed Computing Systems*, (Paker-Verjus Ed.), Academic Press, pp. 67–93.

CERF V.G., KAHN R.E. (1974): A protocol for packet network intercommunication. *IEEE Trans. on Communications*, Vol. C22, 5, 12.

CHANDY K.M. (1985): Concurrent programming for the masses. *Proc. 4th ACM Symposium on Principles of Distributed Computing*, pp. 1–12.

CHANDY K.M., LAMPORT L. (1985): Distributed snapshots: determining global states in distributed systems; *ACM TOCS*, Vol. 3, 1, pp. 63–75.

CHANDY K.M., MISRA J. (1982): Distributed computation on graphs: shortest path algorithms; *Comm. ACM*, Vol. 25, 11, pp. 833–837.

CHANDY K.M., MISRA J. (1985): *A Paradigm for Detecting Quiescent Properties in Distributed Computation*; Springer Verlag, Nato Series F13, (Apt Ed.,), pp. 325–341.

CHANDY K.M., MISRA J. (1986): An example of stepwise refinement of distributed programs: quiescence detection, *ACM TOPLAS*, Vol. 8, 3, pp. 326–343.

CHANDY K.M., MISRA J., HAAS L. (1983): Distributed deadlock detection; *ACM TOCS*, Vol. 1, 2, pp. 144–156.

CHANG E.J.H. (1982): Echo algorithms: depth parallel operations on general graphs. *IEEE Trans. on Soft. Eng.*, Vol. SE 8, 4, pp. 391–401.

CHANG E.J., ROBERTS R. (1979): An improved algorithm for decentralized extrema-finding in circular configurations of processors; *Comm. ACM*, Vol. 22, 5, pp. 281–283.

CHEUNG T.Y. (1983): Graph traversal techniques and the maximum flow problem in distributed computation. *IEEE Trans. on Soft. Eng.*, Vol. SE 9, 4, pp. 504–512.

'CORNAFION' (collective pseudonym) (1981): *Systèmes Informatiques Répartis*, Dunod, 368 pp.

DAHL O.J., DIJKSTRA E.W., HOARE C.A.R. (1972): *Structured Programming*; Academic Press, 220 pp.

DECHTER R., KLEINROCK L. (1986): Broadcast communications and distributed algorithms; *IEEE Trans. on Computers*, Vol. C 35, 3, pp. 210–219.

DIJKSTRA E.W. (1976): *A Discipline of Programming*; Prentice Hall, pp. 217–219.

DIJKSTRA E.W., FEIJEN W.H.J., Von GASTEREN A.J.M. (1983):

Derivation of a termination detection algorithm for distributed computations; *Inf. Processing Letters,* Vol. **16**, pp. 217–219.

DIJKSTRA E. W., SCHOLTEN C. S. (1980): Termination detection for diffusing computations. *Inf. Processing Letters*, Vol. **11**, 1, pp. 217–219.

DOLEV D. (1982): The Byzantine Generals strike again; *Journal of Algorithms*, Vol. **3**, pp. 14–30.

DOLEV D., KLAWE M., RODEH M. (1982): An $O(n \, logn)$ unidirectional distributed algorithm for extrema-finding in a circle. *Journal of Algorithms*, Vol. **3**, pp. 245–260.

DOLEV D., STRONG H.R. (1983): Authenticated algorithms for Byzantine agreement. *SIAM Journal of Computer*, Vol. **12**, 4, pp. 656–666.

EVEN S., *Graph Algorithms*. Pitman, (1979), 249 pp.

FILMAN R.E., FRIEDMAN, D.P. (1984): *Coordinated Computing: tools and techniques for Distributed Software*; McGraw Hill, 370 pp.

FISCHER H.J., LYNCH N.A., PATERSON M.D. (1985): Impossibility of distributed consensus with one faulty process; *Journal of ACM*, Vol. **32**, 2, pp. 374–382.

FRANKLIN W.R. (1982): On an improved algorithm for decentralized extrema-finding in a circular configuration of processors; *Comm. ACM*, Vol. **25**, 5, pp. 336–337.

GALLAGER R.G., HUMBLET P.A., SPIRA P.M. (1983): A distributed algorithm for minimum-weight spanning trees. *ACM Toplas*, Vol. **5**, 1, pp. 66–67.

GARCIA MOLINA H. (1981): Elections in a distributed computing system; *IEEE Trans. on Computers*, Vol. **C 31**, 1, pp. 48–59.

GEHANI N. (1984): Broadcasting sequential processes. *IEEE Trans. on Soft. Eng.*, Vol. **SE 4**, pp. 343–351.

GONDRAN M., MINOUX M. (1979): *Graphes et Algorithmes*; Eyrolles, 518 pp.

GOPAL I.S. (1985): Prevention of store and forward deadlock in computer networks. *IEEE Trans. on Comm.*, Vol. **COM 33**, 12, pp. 1258–1264.

GRIES D. (1981): *The Science of Programming*; Springer Verlag, 366 pp.

GUNTHER K.D. (1981) Prevention of deadlocks in packet-switched data transport systems; *IEEE Trans. on Comm.*, Vol. **C 29**, 4, pp. 512–524.

HAVENDER J.W. (1968): Avoiding deadlocks in multitasking systems; *IBM System Journal*, Vol. **7**, 2, pp. 74–84.

HELARY J.M., MADDI A., RAYNAL M. (1986): *Controlling Knowledge Transfers in Distributed Algorithms, Application to deadlock Detection*; Rapport de Recherche 493, INRIA, 28 pp. Int. Conf. on Parallel Processing, L'aquila, Italy (1987).

HELARY J.M., MADDI A., RAYNAL M. (1987): *Calcul distribué d'un Extremum et du Routage associé dans un Réseau Quelconque*; Rapport de recherche 516, INRIA, (Avril 1986), 36 p. Also in *Revue RAIRO Informatique Théorique et Applications* (1987).

HELARY J.M., PLOUZEAU N., RAYNAL M. (1986): *A distributed Algorithm for Mutual Exclusion in an Arbitrary Network*; Rapport de

Recherche INRIA N° 486, 15 pp. To appear in *Computer Journal* (1988).

HERMAN D. (1981): Contrôle réparti des synchronisations entre processus; *Proc. 2nd Int. Conf. on DC*, Paris, pp. 24–30.

HIRSCHBERG O.S., SINCLAIR J.B. (1980): Decentralized extrema-finding in circular configurations of processors. *Comm. ACM*, Vol. **23**, 11, pp. 627–628.

HOARE C.A.R. (1985): *Communicating Sequential Processes*; Prentice Hall, 256 pp.

HUMBLET P.A. (1983): A distributed algorithm for minimum weight directed spanning tree; *IEEE Trans. on Communications*, Vol. **COM 31**, 6, pp. 756–762.

KORACH E., ROTEM D., SANTORO N. (1984): Distributed algorithms for finding centers and medians in networks; *ACM Toplas*. Vol. **6**, 3, pp. 380–401.

KRAKOWIAK S. (1985): *Principes des Systèmes d'Exploitation des Ordinateurs*; Dunod, 436 pp.

KUMAR D. (1985): A class of termination detection algorithms for distributed computations; *Proc. of Foundations of Soft. Technology*, LNCS n° 206, Springer-Verlag, pp. 73–100.

LAKSHMAN T.V., AGRAWALA A.K. (1986): Efficient decentralized consensus protocols; *IEEE Trans. on Soft. Eng.*, vol. **SE 12**, 5, pp. 600–607.

LAMPORT L. (1974): A new solution of Dijkstra's concurrent programming problem; *Comm. ACM*, Vol. **17**, 8, pp. 453–455.

LAMPORT L. (1978): Time, clocks and the ordering of events in a distributed system; *Comm. ACM*, Vol. **21**, 7, pp. 558–565.

LAMPORT L. (1982): An assertional correctness proof of a distributed algorithm; *Science of Computer Programming*, Vol. **2**, pp. 175–206.

LAMPORT L. (1984): Solved problems, unsolved problems and non-problems in concurrency; *Proc. 3rd ACM Conf. on Principles of Dist. Computing* Reprinted in *ACM O.S. Review*, Vol. **19**, 4, (October 1985), pp. 34–44.

LAMPORT L. (1984): Using time instead of time-out for fault-tolerant distributed systems; *ACM Toplas*, Vol. **6**, 2, pp. 254–280.

LAMPORT L., SCHNEIDER F.B. (1985): Paradigms for distributed programs. in *Distributed Systems* LNCS 190, Springer-Verlag, pp. 431–480.

LAMPORT L., SHOSTAK R., PEASE M. (1982): The Byzantine general problem. *ACM Toplas*, Vol. **4**, 3, pp. 382–401.

LAVALLEE I., ROUCAIROL G. (1986): A fully distributed minimal spanning tree algorithm; *Inf. Processing Letters*, Vol. **23**, pp. 55–62.

LE LANN G. (1977): Distributed systems: towards a formal approach; *IFIP Congress*, Toronto, pp. 155–160.

'LORRAINS' (collective pseudonym) (1979): *Réseaux Télé-informatiques*; Hachette, 303 pp.

MACCHI C., GUILBERT J.F. (1979): *Télé-informatique*; Dunod, 642 pp.

MacCURLEY R., SCHNEIDER F.B. (1986): Derivation of a distributed algorithm for finding paths in directed networks; *Science of Computer Programming*, Vol. **6**, pp. 1–9.

MAEKAWA M. (1985): An O $(\bar{V}n)$ algorithm for mutual exclusion in decentralized systems; *ACM TOCS*, Vol. **3**, 2, pp. 145–159.

MARTIN A.J. (1985): Distributed mutual exclusion on a ring of processes. *Science of Computer Programming*, Vol. **5**, pp. 256–276.

MERLIN P.M. (1979): Specification and validation of protocols; *IEEE Trans. on Communications*, Vol. **COM 27**, 11, pp. 1671–1681.

MERLIN P.M., SCHWEITZER P.J. (1980): Deadlock avoidance in store and forward networks; *IEEE Trans. on Comm.*, Vol. **COM 28**, 3, pp. 345–354.

MERLIN P.M., SEGALL A. (1979): A failsafe distributed routing protocol; *IEEE Trans. on Communications*. Vol. **COM 27**, 9, pp. 1280–1288.

MISRA J., CHANDY K.M. (1982): A distributed graph algorithm: knot detection; *ACM Toplas*, Vol. **4**, 4 pp. 678–686.

MISRA J. (1983): Detecting termination of distributed computations using markers; *Proc. 2nd ACM Conf. on Principles of DC*, Montréal, pp. 290–294.

MORGAN C. (1985): Global and logical time in distributed algorithms; *Inf. Processing Letters*, Vol. **20**, (1985), pp. 189–194.

MUNTEAN T. (1977): Formalism for the specification of unities controlling Process Cooperation: *Proc. of Conf. on Inf. Sciences and Systems*, Baltimore.

NATARAJAN N. (1986): A distributed scheme for detecting communication deadlocks; *IEEE Trans. on Soft. Eng.*, Vol. **SE 12**, 4, pp. 531–537.

PACH J.A., KORACH E., ROTEM D. (1984): Lower bounds for distributed maximum-finding algorithms; *Journal of the ACM*, Vol. **31**, 4, pp. 905–918.

PERRY K.J., TOUEG S. (1986): Distributed agreement in the presence of processor and communication faults; *IEEE Trans. on Soft. Eng.*, Vol. **SE 12**, 3, pp. 477–481.

PETERSON G.L. (1982): An $O(n \log n)$ unidirectional algorithm for the circular extrema problem; *ACM Toplas*, Vol. **4**, 4, pp. 758–762.

PETERSON J.L., SILBERSCHATZ A. (1983): *Operating Systems Concepts*; Addison-Wesley, 548 pp.

PLOUZEAU N., RAYNAL M., VERJUS J.P. (1987): Producteurs-consommateur: quelques solutions réparties; *TSI*, Vol. **6**, 3.

PRICE W.L. (1974): Simulation studies of an isarithmically controlled store and forward data communication network; *IFIP Congress*, pp. 151–154.

PUJOLLE G., SERET D., DROMMARD D., HORLAIT E. (1985): *Réseaux et Télématique*; Eyrolles, Tome 1: 315 pp. Tome 2: 330 pp.

RANA S.P. (1983): A distributed solution of the distributed termination problem; *Inf. Processing Letters*, Vol. **17**, pp. 43–46.

RAYNAL M. (1983): Une analyse de la spécification de la coopération entre processus par variables partagées; *TSI*, Vol. **1**, 3, pp. 201–211.

RAYNAL M. (1986): *Algorithms for Mutual Exclusion*; North Oxford Academic, 107 pp.

RAYNAL M. (1987): *Distributed Algorithms and Protocols*; John Wiley.

RAYNAL M. (1985): Un algorithme d'exclusion mutuelle pour une structure logique en anneau. *TSI*, Vol. **4**, 5, pp. 471–474.

RAYNAL M. (1987): A distributed algorithm to prevent mutual drift between n logical clocks; *Inf. Processing Letters*, Vol. 24, 1987; pp. 199–202.

RAYNAL M., RUBINO G. (1985): *Détecter la perte de jetons et les Regénérer sur une Structure en Anneau*; Rapport de Recherche INRIA, no. 428, 24 pp. Also in Proc. Int. Conf. on Parallel Processing, L'Aquila, Italy (1987).

RICART G., AGRAWALA A.K. (1981): An optimal algorithm for mutual exclusion in computer networks; *Comm. ACM*, Vol. **24**, 1, pp. 9–17.

RICART G., AGRAWALA A.K. (1983): Author response to "On mutual exclusion in computer networks" by Carvalho and Roucairol; *Comm. ACM*, Vol. 26, 2, pp. 147–148.

ROBERT P., VERJUS J.P. (1977): Towards autonomous descriptions of synchronisation modules; *IFIP Congress*, Toronto, pp. 981–986.

ROTEM D., SANTORO N., SIDNEY J., (1985): Distributed sorting; *IEEE Trans. on Computers*, Vol. **C 34**, 4, pp. 372–376.

SCHIPER A. (1986): *Programmation Concurrente*; Presses Polytechniques Romandes Lausanne, 295 pp.

SCHNEIDER F.B., GRIES D., SCHLICHTING R.O. (1984): Fault tolerant broadcasts; *Science of Computer Programming*, Vol. **4**, 1, pp. 1–15.

SEGALL A. (1982) Decentralized maximum flow protocols; *Networks*, Vol. **12**, pp. 213–230.

SEGALL A. (1985): Distributed networks protocols; *IEEE Trans. on Information Theory*, Vol. **IT 29**, 1, pp. 23–35.

SEITZ C. (1985): The cosmic cube. *Comm. ACM*. Vol. **28**, 1, pp. 22–33.

STENNING W. (1976): A data transfer protocol; *Computer Networks*, Vol. **1**, (1976), pp. 99–110.

SUZUKI I., KASAMI T. (1982): On optimality theory for mutual exclusion algorithms in computer networks; *Proc. 3rd Int. Conf. on DC Systems*, Miami, pp. 367–370.

TAJIBNAPIS W.D. (1977): A correctness proof of a topology information maintenance protocol for distributed computer networks; *Comm. of the ACM*, Vol. **20**, 7, pp. 477–485.

TANENBAUM A.S. (1981): *Computer Networks*; Prentice Hall, (1981), 518 pp.

TRAIGER I.L., GRAY J., GALTIERI C.A., LINDSAY B.G. (1982): Transactions and consistency in distributed database systems; *ACM TODS*, Vol. **7**, 3, pp. 323–342.

TREHEL M., NAIMI M. (1986): Un algorithme d'exclusion mutuelle dans un système distribué en O(logn). Rapport de Recherche, Université de Besançon, (*TSI*, Vol. 6, 2; 1987).

ULLMAN J.D. (1982): *Principles of Data Base Systems*; 2nd Edition Computer Science Press, 484 pp.

VERJUS J.P. (1983): Synchronization in distributed systems: an informal introduction; in Distributed Computing Systems, (Paker-Verjus Ed.), Academic Press, pp. 3–22.

VERJUS J.P., THOROVAL R. (1986): Dérivation d'algorithmes distribués d'arbitrage; *TSI*, Vol. 5, 1, pp. 37–48.

WUU G.T., BERNSTEIN A. (1985): False deadlock detection in distributed systems; *IEEE Trans. on Soft. Eng.*, Vol. **SE 11**, 8, pp. 820–821.

ZACKS S. (1985): Optimal distributed algorithms for sorting and ranking; *IEEE Trans. on Computers*, Vol. **C 34**, 4, pp. 376–379.

ZIMMERMAN H. (1980): OSI Reference Model— The OSI model of architecture for open systems interconnexion *IEEE Trans. on Comm.*, Vol. **C 28**, pp. 425–432.

Index

The MIT Press, with Peter Denning, general consulting editor, and Brian Randell, European consulting editor, publishes computer science books in the following series:

ACM Doctoral Dissertation Award and Distinguished Dissertation Series

Artificial Intelligence, Patrick Winston and Michael Brady, editors

Charles Babbage Institute Reprint Series for the History of Computing, Martin Campbell-Kelly, editor

Computer Systems, Herb Schwetman, editor

Exploring with Logo, E. Paul Goldenberg, editor

Foundations of Computing, Michael Garey and Albert Meyer, editors

History of Computing, I. Bernard Cohen and William Aspray, editors

Information Systems, Michael Lesk, editor

Logic Programming, Ehud Shapiro, editor; Fernando Pereira, Koichi Furukawa, and D.H.D. Warren, associate editors

The MIT Electrical Engineering and Computer Science Series

Scientific Computation, Dennis Gannon, editor